Mathematical Methods for Chemists

R. K. Mackie
T. M. Shepherd
C. A. Vincent

Department of Chemistry
University of St Andrews

The English Universities Press Limited

ISBN 0 340 08217 8 Boards
ISBN 0 340 08220 8 Paperback

First printed 1972

The English Universities Press Ltd
St Paul's House Warwick Lane London EC4P 4AH

Printed in Great Britain by
Bell and Bain Ltd Glasgow

Preface

The fundamental basis of chemistry as a physical science is not simply the measurement of physical quantities, but the establishment of relationships between these quantities. An exact description of the relationship between the values of one quantity y and those of another quantity x may be given in terms of a mathematical expression, $y = f(x)$. y is said to be 'a function of x'. It is one of the main tasks of both experimental and theoretical chemists to seek such functions or relationships.

At the simplest level, the use of mathematics enables chemists to make precise descriptions of relationships between physical quantities. More fundamentally, however, the methods of mathematics permit relationships to be established that cannot be arrived at intuitively.

What, then, is the proper level of mathematical ability for the student of chemistry? The answer is probably that the higher the level achieved, the more readily will applications of mathematical techniques be made to chemical problems. However, few chemistry students have the time or the desire to bring their mathematics background to a uniformly high level in all the relevant branches of the subject. What does seem important is for students (*a*) to acquire some facility in the commoner mathematical manipulations and (*b*) to cultivate sufficient mathematical ability that the comprehension of subjects which involve more mathematics than is usual in chemistry—such as quantum chemistry, statistical mechanics, spectroscopy, etc.—is not delayed by the novelty or difficulty of the mathematics.

This book is not intended as a substitute for a formal mathematics course. Such a course would involve the study of the following mathematics in much more detail, but, more important, would also enable the student to undertake many more problems and thus be able to *apply* the mathematics to new situations. The object of the book is rather to provide a background knowledge of the mathematics relevant to the chemist, together with an introduction to concepts which may not have been encountered previously. It should also act as a source book for information or definition, and will at least enable its readers to *follow* a considerable range of mathematical descriptions in text-books, research papers, etc.

Contents

1 *Functions and Operators*

1.1 Functions

A mathematical expression involving a variable parameter is known as a *function* of that parameter if its value can be determined when the value of the variable is known. Thus for example

$$2x+3$$

is a function of x. A function can be defined as having validity over all or part of the numerical domain. For instance a function may be defined as valid 'for all real numbers' or 'for all positive integers', etc.

The function of a variable x is usually represented by $\mathrm{f}(x)$, and the function is referred to as y, i.e.

$$y = \mathrm{f}(x)$$

y is then the 'dependent variable' since its value depends on that of x. For example the current passing through a fixed resistor is a function of the applied voltage, E

$$i = \mathrm{f}(E)$$

In fact $i = (1/R)E$ where R is a constant. A function may depend on more than one independent variable: in such a case it is usually written as

$$y = \mathrm{f}(u, v, w, \ldots)$$

where $u, v, w, \ldots$ are variables which are not in any way mathematically related. Thus for instance the pressure of one mole of gas depends on the volume V, and temperature T, and we have

$$P = \mathrm{f}(V, T)$$

For a perfect gas

$$P = \mathrm{f}_1(V, T) = \frac{RT}{V}, \text{ where } R \text{ is a constant.}$$

For an imperfect gas

$$P = \mathrm{f}_2(V, T) = \frac{RT}{V-b} - \frac{a}{V^2}, \text{ where } R, a \text{ and } b \text{ are constants.}$$

Note that $\mathrm{f}_1(V, T)$ and $\mathrm{f}_2(V, T)$ are different functions.

Functions may be described as continuous, discontinuous, periodic, etc. Continuous functions are sometimes known as 'well behaved'.

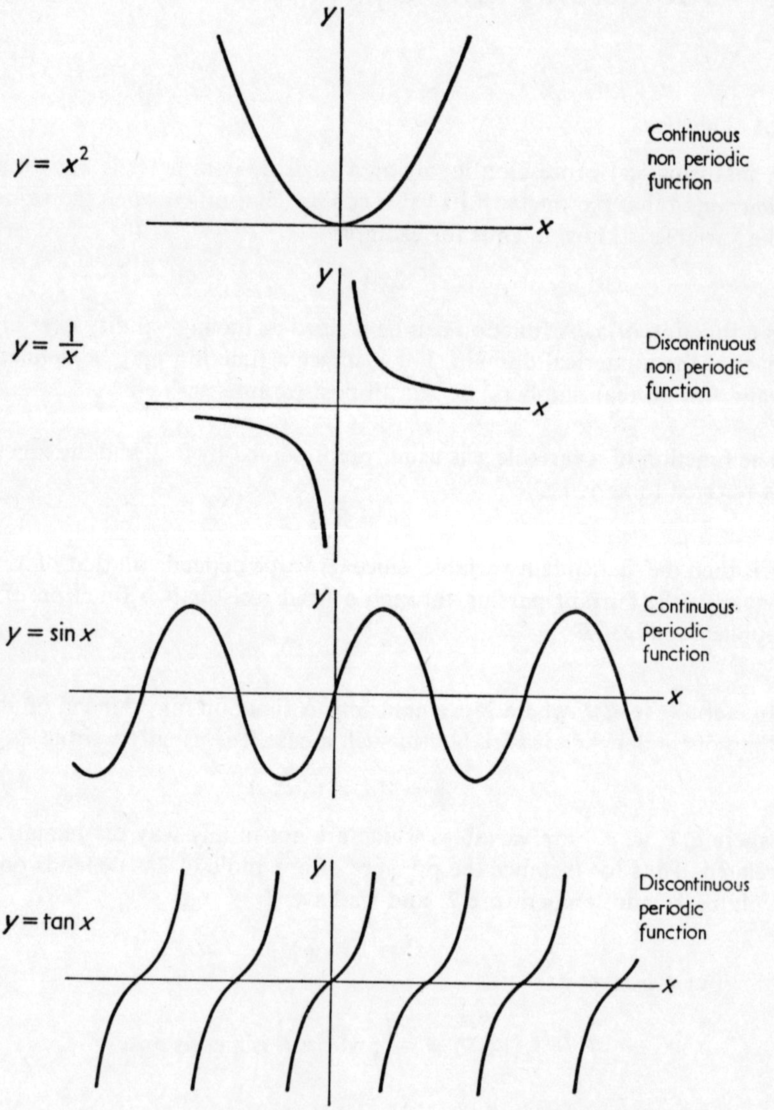

Fig. 1.1

If the function has only one independent variable it may be represented in two dimensions, as in Fig. 1.1.

Functions in two variables may be represented in three dimensions, e.g. the function

$$z = (c^2 - x^2 - y^2)^{\frac{1}{2}}$$

where c is a constant and x and y are independent variables may be represented by the surface of a sphere of radius c (Fig. 1.2).

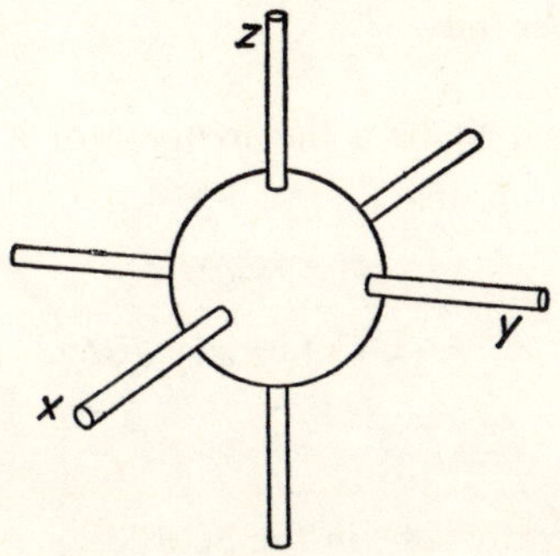

Fig. 1.2

1.2 Operators

Every mathematical operation such as adding, multiplying, etc., may be represented by a symbol characteristic of the operation. For example if we consider 'log y', then the operation required is that of finding the logarithm of the function y and the instruction 'log' is known as a *mathematical operator* denoting the operation to be carried out on the function y. Further examples of operators include $\surd$, sin, exp, $\mathrm{d}/\mathrm{d}x$, $\int \mathrm{d}x$, etc. The function, is known as the *operand.*

More than one operator may operate on a function and it is necessary to be quite specific about the order of operations. If we have two operators, denoted by $\mathscr{A}$ and $\mathscr{B}$, operating on a function, say y, there are two possibilities:

$$\mathscr{A}\mathscr{B}(y) \quad \text{or} \quad \mathscr{B}\mathscr{A}(y)$$

which need not give the same result.

$\mathscr{A}\mathscr{B}(y)$ implies that the operator $\mathscr{B}$ acts on y, followed by $\mathscr{A}$ on the result of the first operation i.e. the innermost operator operates first,

e.g. if $\mathscr{A} = 2$

$\mathscr{B} = \sin$

then $\mathscr{A}\mathscr{B}(y) = 2 \sin y.$

On the other hand $\mathscr{B}\mathscr{A}(y) = \sin 2y$

$$= 2 \cos y \sin y.$$

Clearly in this case $\mathscr{A}\mathscr{B}(y) \neq \mathscr{B}\mathscr{A}(y)$ and the operators are said not to *commute*.

Operators alone have no significance but the algebra of operators may be developed without operands.

The operator $\mathscr{A}$ is said to be a linear operator if it obeys the following relationships:

(i) $$\mathscr{A}cx = c\mathscr{A}x$$

and (ii) $$\mathscr{A}(x+y) = \mathscr{A}x+\mathscr{A}y$$

where c is a constant.

The Hamiltonian operator, $\mathscr{H}$, in the Schrödinger equation,

$$\mathscr{H}\Psi = E\Psi,$$

is an example of a linear operator. $\mathscr{H}$ acts on the eigenfunction (or wave function) Ψ giving a result which is the product of the energy of the system, E (i.e. the eigenvalue) and the eigenfunction itself. *It is imperative to remember that $\mathscr{H}$ is an operator and E simply a number.* Ψ is generally a function of position and time.

2 *Differential Calculus*

In many areas of chemistry we are interested in continuous changes taking place in a system. Differential calculus describes the rate at which such changes take place.

2.1 Rate of change of the function $y = x^2$

Let us consider the value of y at various values of x

x	1	2	5	6	10	11
$y = x^2$	1	4	25	36	100	121

The difference between pairs of values of x, i.e. Δx, is 1, while values of Δy, the difference between the corresponding pairs of values of y, are 3, 11, and 21 respectively.

If we take the rate of change of the function y with respect to x to be $\Delta y/\Delta x$, then we see that this parameter changes with the value of x,

i.e.
$$\left(\frac{\Delta y}{\Delta x}\right)_1 = \frac{3}{1} = 3$$
$$\left(\frac{\Delta y}{\Delta x}\right)_2 = \frac{11}{1} = 11$$
$$\left(\frac{\Delta y}{\Delta x}\right)_3 = \frac{21}{1} = 21.$$

Let us now consider more closely the behaviour of this rate of change of the function about the value $x = 5$.

x	5	5·01	5·1	6
$y = x^2$	25	25·1001	26·01	36
Δy	0	0·1001	1·01	11
Δx	0	0·01	0·1	1
$\Delta y/\Delta x$	indeterminate	10·01	10·1	11

and in general at $x = 5+h$, $\quad y = 25+10h+h^2$

$$\Delta y = 10h+h^2$$

$$\Delta x = h$$

$$\Delta y/\Delta x = 10+h.$$

dy/dx is defined as the limit of $\Delta y/\Delta x$ as h tends to zero,

i.e. $$\lim_{h\to 0} \frac{\Delta y}{\Delta x} = dy/dx = 10 \quad \text{when } x = 5.$$

2.2 Tangents to curves

Fig. 2.1 shows the graph of $y = x^2$.

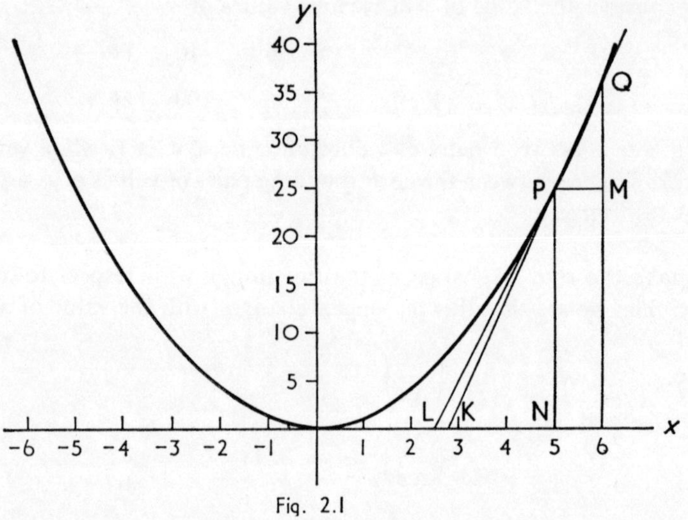

Fig. 2.1

If P is the point (5, 25) on the graph and Q the point (6, 36) then the increase in y, QM, corresponds to the increase in x, PM, and the rate of increase of the function

$$= \Delta y/\Delta x$$

$$= \text{QM/PM} = 11$$

$$= \tan \angle\text{QPM} = \tan \angle\text{PKN}.$$

In the limit when the increase in x is very small (i.e. $\Delta y/\Delta x \to dy/dx$) the chord QP becomes the tangent to the curve at P, PL. Thus, dy/dx would equal tan $\angle$PLN if the graph were isometric (Section 11.2) and, since we have discovered in 2.1 the value of dy/dx to be 10 when $x = 5$, from tables $\angle\text{PLN} = 84° \, 18'$.

2.3 Differentiation from first principles

The process of finding the limit of $\Delta y/\Delta x$ as the increments in x tend to zero (i.e. evaluating $\mathrm{d}y/\mathrm{d}x$) is called *differentiation* and $\mathrm{d}y/\mathrm{d}x$ is called the *differential coefficient* of y with respect to x. It is also known as the *derived function* or *derivative*. Let us differentiate the function $y = x^2$.

If y changes to $y+\Delta y$ as x changes to $x+\Delta x$, then

$$y+\Delta y = (x+\Delta x)^2$$

i.e. $$y+\Delta y = x^2+2x\Delta x+(\Delta x)^2$$

Now $$y = x^2$$

therefore $$x^2+\Delta y = x^2+2x\Delta x+(\Delta x)^2$$

i.e. $$\Delta y = 2x\Delta x+(\Delta x)^2$$

Dividing by Δx

$$\Delta y/\Delta x = 2x+\Delta x.$$

In the limit when Δx tends to 0, $\Delta y/\Delta x$ becomes $\mathrm{d}y/\mathrm{d}x$

and thus $$\mathrm{d}y/\mathrm{d}x = 2x.$$

This can be applied to more complex functions

e.g. $$y = \frac{1}{2x+1}$$

$$y+\Delta y = \frac{1}{2(x+\Delta x)+1}$$

Substituting $$y = \frac{1}{2x+1}$$

we have $$\frac{1}{2x+1}+\Delta y = \frac{1}{2x+2\Delta x+1}$$

Rearranging, $$\Delta y = \frac{1}{2x+2\Delta x+1}-\frac{1}{2x+1}$$

$$= \frac{2x+1-2x-2\Delta x-1}{(2x+2\Delta x+1)(2x+1)}$$

$$= \frac{-2\Delta x}{(2x+2\Delta x+1)(2x+1)}$$

Dividing by Δx

$$\Delta y/\Delta x = \frac{-2}{(2x+2\Delta x+1)(2x+1)}$$

so that

$$\lim_{\Delta x \to 0} \frac{\Delta y}{\Delta x} = \mathrm{d}y/\mathrm{d}x = \frac{-2}{(2x+0+1)(2x+1)}$$

$$= \frac{-2}{(2x+1)^2}$$

Use of methods analogous to these enables us to evaluate the differential coefficients listed in Appendix B.

2.4 Differentiation of a function of a function

If u is a function of x

then $\mathrm{d}y/\mathrm{d}x = \mathrm{d}y/\mathrm{d}u . \mathrm{d}u/\mathrm{d}x$

Example. Differentiate $\sqrt{(x^2+1)}$

For $y = \sqrt{(x^2+1)}$ put $u = x^2+1$

Now y is a function of u, ($y = \sqrt{u}$), which in turn is a function of x, ($u = x^2+1$), i.e. y is a function of a function.

Now $y = \sqrt{u} = u^{\frac{1}{2}}$

therefore $\mathrm{d}y/\mathrm{d}u = \frac{1}{2}u^{-\frac{1}{2}} = \dfrac{1}{2\sqrt{u}}$

and $u = x^2+1$

therefore $\mathrm{d}u/\mathrm{d}x = 2x$

and

$$\mathrm{d}y/\mathrm{d}x = \mathrm{d}y/\mathrm{d}u . \mathrm{d}u/\mathrm{d}x$$

$$= \frac{1}{2\sqrt{u}} 2x$$

$$= \frac{2x}{2\sqrt{(x^2+1)}} = \frac{x}{\sqrt{(x^2+1)}}$$

N.B. If $u, v, w, \ldots z$ are functions of x then

$$\mathrm{d}y/\mathrm{d}x = \mathrm{d}y/\mathrm{d}u . \mathrm{d}u/\mathrm{d}v . \mathrm{d}v/\mathrm{d}w . \ldots \mathrm{d}z/\mathrm{d}x.$$

2.5 Alternative notations

We have used the notation $\mathrm{d}y/\mathrm{d}x$ for the differential coefficient of $y = \mathrm{f}(x)$.

Other notations in use are y', $\mathrm{f}'(x)$, $D_x y$ and $\mathrm{d}/\mathrm{d}x\ \mathrm{f}(x)$.

2.6 Logarithmic differentiation

If $y = u^v$ where u and v are both functions of x it is convenient to take logs before differentiating

$$\ln y = v \ln u$$

Differentiating both sides with respect to x, and since from the list of standard forms $\mathrm{d}/\mathrm{d}x(\ln x) = 1/x$

and $\mathrm{d}/\mathrm{d}x(u.v) = u\,\mathrm{d}v/\mathrm{d}x + v\,\mathrm{d}u/\mathrm{d}x$

we have $1/y.\mathrm{d}y/\mathrm{d}x = \ln u\,\mathrm{d}v/\mathrm{d}x + v/u\,\mathrm{d}u/\mathrm{d}x$

i.e.

$$\mathrm{d}y/\mathrm{d}x = y[\ln u\,\mathrm{d}v/\mathrm{d}x + v/u\,\mathrm{d}u/\mathrm{d}x]$$

$$= u^v[\ln u\,\mathrm{d}v/\mathrm{d}x + v/u\,\mathrm{d}u/\mathrm{d}x]$$

The function $y = c^x$ can also be differentiated in this way.

Taking logarithms, $\ln y = x \ln c$

Using the symbolism of the first part of this section

$u = \ln c$, a constant, therefore $\mathrm{d}u/\mathrm{d}x = 0$

$v = x$, therefore $\mathrm{d}v/\mathrm{d}x = 1$

Substituting in the equation derived above

$$\mathrm{d}y/\mathrm{d}x = c^x[\ln c.1+0]$$

$$= c^x \ln c$$

2.7 Simplification of relationships by taking logarithms

It is sometimes found that a simpler relationship between two variables can be derived by use of logarithms.

The following two examples show instances in which this technique can be used.

Example. The repulsive force, ϕ, between two molecules of gas at a distance a from each other is given by:

$$\phi = Aa^{-n} \quad \text{where } A \text{ and } n \text{ are constants.}$$

Show how the value of n may be obtained from a study of the temperature dependence of the viscosity of the gas.

When two molecules approach each other as close as possible, there is no relative motion between them and their kinetic energy is converted into potential energy. We then have

$$kT = A\sigma^{-n}$$

where σ is the value of a when the molecules are at the distance of closest approach

i.e. $$\sigma = (A/kT)^{1/n} \qquad \text{...(2A)}$$

Now, the viscosity of the gas is given by

$$\eta = \frac{1}{\pi\sigma^2}\left(\frac{mkT}{\pi}\right)^{\frac{1}{2}} \qquad \text{...(2B)}$$

where m is the mass of the molecule.

Experimentally it is found that the variation of viscosity with temperature is given by

$$\eta = \text{constant} \times T^c \qquad \text{...(2C)}$$

We can now evaluate n in terms of c.

Taking logs of equation (2B) we have

$$\ln \eta = \ln \frac{1}{\pi} - 2 \ln \sigma + \tfrac{1}{2}\left(\ln \frac{mk}{\pi} + \ln T\right)$$

Differentiating with respect to $\ln T$

$$\frac{d \ln \eta}{d \ln T} = -2\frac{d \ln \sigma}{d \ln T} + \tfrac{1}{2}(0+1)$$

Taking logs of equation (2A) and then differentiating with respect to $\ln T$

$$\ln \sigma = \frac{1}{n}(\ln A/k - \ln T)$$

therefore $$\frac{d \ln \sigma}{d \ln T} = -\frac{1}{n}$$

therefore $$\frac{d \ln \eta}{d \ln T} = \tfrac{1}{2} + \frac{2}{n} \qquad \text{...(2D)}$$

From the experimental equation (2C), we obtain

$$\ln \eta = \ln(\text{constant}) + c \ln T \qquad \text{...(2E)}$$

i.e. $$\frac{d \ln \eta}{d \ln T} = c$$

Equating the expressions (2D) and (2E) we find that

$$c = \tfrac{1}{2}+\frac{2}{n}$$

i.e.
$$c-\tfrac{1}{2} = \frac{2c-1}{2} = \frac{2}{n}$$

therefore
$$n = \frac{4}{2c-1}$$

Example. The equilibrium constant expressed in terms of mole fraction, K_x, is pressure dependent while the equilibrium constant in terms of partial pressure, K_p, is independent of pressure. From the relationship

$$K_x = K_p p^{-\Delta n}$$

show that K_x decreases with increasing pressure, p, if Δn, the number of molecules of products less that of reactants in the stoichiometric equation for the reaction, is greater than zero.

To differentiate this equation, it is convenient to take logarithms.

$$\ln K_x = \ln K_p + \ln p^{-\Delta n}$$
$$= \ln K_p - \Delta n \ln p$$

therefore
$$\frac{\mathrm{d} \ln K_x}{\mathrm{d}p} = 0-\Delta n\left(\frac{1}{p}\right) = \frac{-\Delta n}{p}$$

Thus if Δn is positive, $(\mathrm{d} \ln K_x/\mathrm{d}p)$ is negative, i.e. K_x decreases with increasing pressure.

e.g. for $N_2O_4 \rightleftharpoons 2NO_2$, $\Delta n = 1$

and in this case the equilibrium constant will decrease as the pressure of N_2O_4 is increased.

2.8 Partial differentiation

If z is a function of other independent variables as well as x, the derivative of z with respect to x is called the *partial derivative* of z with respect to x. It can be interpreted as the rate of change of z with respect to x while all the other variables remain constant. It is written $\partial z/\partial x$ or $(\partial z/\partial x)_{u,v,w}$, if we wish to draw attention to the other variables, u, v and w, of which z is a function.

As can be seen from Figs. 2.2(a) and 2.2(b), $z = f(x, y)$ describes a surface. $\partial z/\partial x$ at a point, P, on the surface is the tangent at P to the curve which is formed by the plane through P parallel to the xz plane cutting the surface.

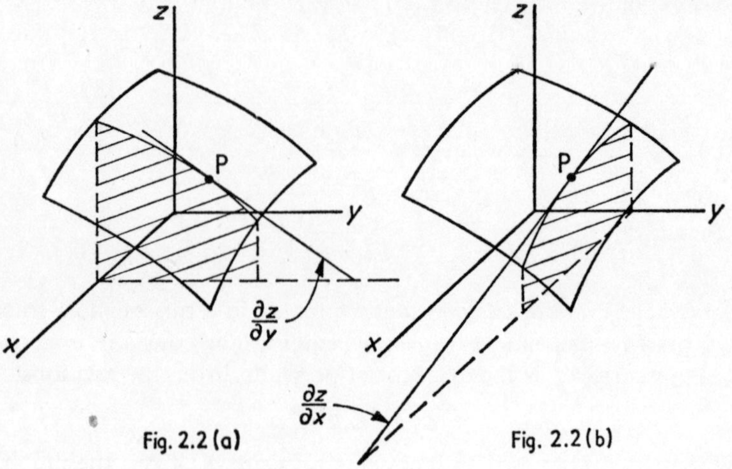

Fig. 2.2 (a) Fig. 2.2 (b)

Similarly $\partial z/\partial y$ is the tangent to the curve which is formed by the plane through P parallel to the yz plane cutting the surface.

Example. For an ideal gas $PV = nRT$, i.e. the pressure depends on the independent variables V and T.

$$P = \frac{nRT}{V}$$

$$\frac{\partial P}{\partial T} = \left(\frac{\partial P}{\partial T}\right)_V = \frac{nR}{V}$$

$$\frac{\partial P}{\partial V} = \left(\frac{\partial P}{\partial V}\right)_T = \frac{-nRT}{V^2}$$

These equations now give us respectively the variation of pressure with temperature at constant volume and the variation of pressure with volume at constant temperature.

2.9 Differentials

We have up to this stage considered the differential coefficient $\mathrm{d}y/\mathrm{d}x$ as an entity. However, in some circumstances it is useful to separate the variables x and y. To do this we define *differentials* $\mathrm{d}x$ and $\mathrm{d}y$ such that dividing $\mathrm{d}y$ by $\mathrm{d}x$ we get the differential coefficient $\mathrm{d}y/\mathrm{d}x$.

Let
$$y = ax^3$$
$$\mathrm{d}y/\mathrm{d}x = 3ax^2$$

We can now rewrite this equation, using differentials, as

$$\mathrm{d}y = 3ax^2\,\mathrm{d}x$$

dy and dx can be regarded as infinitely small changes in y and x. We will make use of this manipulation during our discussion of integration.

2.10 Total differential of a function of two variables

If z is a function of two independent variables x and y,

then
$$\mathrm{d}z = \frac{\partial z}{\partial x}\,\mathrm{d}x + \frac{\partial z}{\partial y}\,\mathrm{d}y$$

dz is known as the *total* (or *exact* or *complete*) *differential* of z; dx and dy are differentials. This relationship enables us to calculate the change in z due to simultaneous changes in x and y. A simple geometrical example may help to clarify this procedure.

The area, A, of rectangle BCDE whose sides are x and y is given by

$$A = xy$$

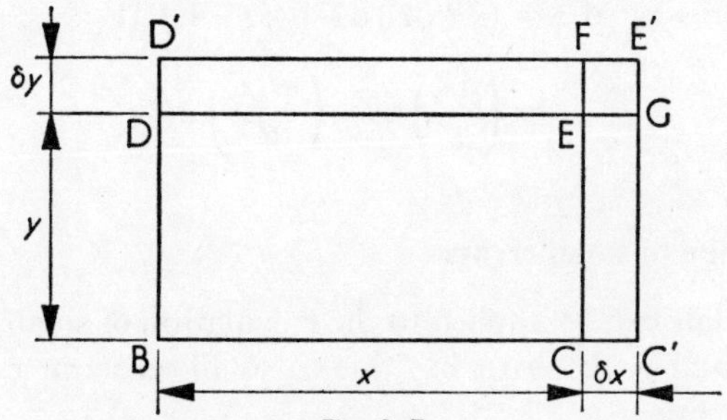

Fig. 2.3

$$\partial A/\partial x = \lim_{\delta x \to 0} \frac{\text{increase in area}}{\text{increase in length}}, \quad y \text{ remaining constant}$$

$$= \lim_{\delta x \to 0} \frac{\mathrm{EC.CC'}}{\mathrm{CC'}} = \mathrm{EC} = y$$

$$\partial A/\partial y = \lim_{\delta y \to 0} \frac{\text{increase in area}}{\text{increase in width}}, \quad x \text{ remaining constant}$$

$$= \lim_{\delta y \to 0} \frac{\mathrm{ED.DD'}}{\mathrm{DD'}} = \mathrm{ED} = x$$

If the length is increased from BC to BC′ simultaneously with the width being increased from BD to BD′, the increase in area is δA

$$= \mathrm{DD'.DE+CC'.CE+EG.EF}$$

$$= x\delta y + y\delta x + \delta x . \delta y.$$

$$\mathrm{d}A = \lim_{\delta x,\, \delta y \to 0} \delta A = x\,\mathrm{d}y + y\,\mathrm{d}x$$

(dx and dy are limits of δx and δy)

since $\delta x . \delta y \to 0$ as δx, $\delta y \to 0$.

The total differential of $A = xy$ is given by

$$\mathrm{d}A = \frac{\partial A}{\partial x}\,\mathrm{d}x + \frac{\partial A}{\partial y}\,\mathrm{d}y$$

If $A = xy$, then $\mathrm{d}A = y\,\mathrm{d}x + x\,\mathrm{d}y$ which is the result obtained geometrically.

Example. Calculate the change in pressure in an ideal gas due to simultaneous changes in volume and temperature.

$$PV = nRT$$

i.e.

$$P = nRT/V$$

$$\mathrm{d}P = (\partial P/\partial T)\,\mathrm{d}T + (\partial P/\partial V)\,\mathrm{d}V$$

$$= \left(\frac{nR}{V}\right)\mathrm{d}T - \left(\frac{nRT}{V^2}\right)\mathrm{d}V$$

2.11 Application to small errors

Total differentials can be applied to the calculation of small errors. If f is a function of x and y, the error in f due to small errors in x and y is given approximately by

$$\delta f \approx (\partial f/\partial x)\delta x + (\partial f/\partial y)\delta y$$

Example. According to Henry's law the solubility of a gas is proportional to its partial pressure. The volume of a perfect gas dissolving is then related to the volume dissolving under standard conditions by the expression

$$\frac{v}{v_0} = \frac{p_0}{p}\frac{T}{T_0}$$

Calculate the error in the volume of gas dissolving, δv, due to errors in the measurement of p and T of δp and δT respectively.

$$\delta v = \left(\frac{\partial v}{\partial T}\right)\delta T + \left(\frac{\partial v}{\partial p}\right)\delta p$$

$$= \left(\frac{p_0 v_0}{pT_0}\right)\delta T + \left(\frac{p_0 v_0 T}{T_0}\left(-\frac{1}{p^2}\right)\right)\delta p$$

$$= \frac{p_0 v_0 T}{pT_0}\left(\frac{\delta T}{T} - \frac{\delta p}{p}\right)$$

$$= \left(\frac{\delta T}{T} - \frac{\delta p}{p}\right) v$$

Note that in this case if the relative errors in p and T are equal i.e. say both are equal to $+1\%$, δv is zero.

2.12 Thermodynamic relationships using partial derivatives

It can be shown that if z is a continuous function of x and y

$$\frac{\partial^2 z}{\partial y \partial x} = \frac{\partial^2 z}{\partial x \partial y}$$

i.e.

$$\frac{\partial}{\partial y}\left(\frac{\partial z}{\partial x}\right) = \frac{\partial}{\partial x}\left(\frac{\partial z}{\partial y}\right)$$

This is sometimes known as the *cross-derivative* relationship and it can be used for example in thermodynamics to derive the Maxwell relationships. Starting with the fundamental equation

$$\mathrm{d}U = T\,\mathrm{d}S - P\,\mathrm{d}V$$

we can equate this with the complete derivative of $U(S, V)$

$$\mathrm{d}U = \left(\frac{\partial U}{\partial S}\right)_V \mathrm{d}S + \left(\frac{\partial U}{\partial V}\right)_S \mathrm{d}V$$

Thus

$$T = \left(\frac{\partial U}{\partial S}\right)_V \quad \text{and} \quad -P = \left(\frac{\partial U}{\partial V}\right)_S$$

Using the cross-derivative relationship for U as a function of S and V

$$\frac{\partial}{\partial V}\left(\frac{\partial U}{\partial S}\right) = \frac{\partial}{\partial S}\left(\frac{\partial U}{\partial V}\right)$$

we obtain

$$\frac{\partial}{\partial V}(T) = \frac{\partial}{\partial S}(-P)$$

i.e.

$$\left(\frac{\partial T}{\partial V}\right)_S = -\left(\frac{\partial P}{\partial S}\right)_V$$

This is the first of the Maxwell relationships and we can obtain the others in analogous ways.

$$\left(\frac{\partial T}{\partial P}\right)_S = \left(\frac{\partial V}{\partial S}\right)_P \quad \text{from} \quad \mathrm{d}H = T\,\mathrm{d}S + V\,\mathrm{d}P$$

$$\left(\frac{\partial S}{\partial V}\right)_T = \left(\frac{\partial P}{\partial T}\right)_V \quad \text{from} \quad \mathrm{d}A = -S\,\mathrm{d}T - P\,\mathrm{d}V$$

and $$\left(\frac{\partial S}{\partial P}\right)_T = -\left(\frac{\partial V}{\partial T}\right)_P \quad \text{from} \quad \mathrm{d}G = -S\,\mathrm{d}T + V\,\mathrm{d}P$$

Another useful identity is the *cyclic relationship*

$$\left(\frac{\partial x}{\partial y}\right)_z \left(\frac{\partial y}{\partial z}\right)_x \left(\frac{\partial z}{\partial x}\right)_y = -1$$

If z is a function of x and y then its complete differential is

$$\mathrm{d}z = \left(\frac{\partial z}{\partial x}\right)_y \mathrm{d}x + \left(\frac{\partial z}{\partial y}\right)_x \mathrm{d}y$$

If we now restrict the system to those changes in x and in y which do not alter z, i.e. $\mathrm{d}z = 0$, $\mathrm{d}x$ and $\mathrm{d}y$ become $(\partial x)_z$ and $(\partial y)_z$ respectively, and then

$$0 = \left(\frac{\partial z}{\partial x}\right)_y (\partial x)_z + \left(\frac{\partial z}{\partial y}\right)_x (\partial y)_z$$

Dividing by $(\partial y)_z$

$$0 = \left(\frac{\partial z}{\partial x}\right)_y \left(\frac{\partial x}{\partial y}\right)_z + \left(\frac{\partial z}{\partial y}\right)_x \quad \text{since} \quad \left(\frac{\partial y}{\partial y}\right)_z = 1$$

Multiplying by $(\partial y/\partial z)_x$

$$0 = \left(\frac{\partial z}{\partial x}\right)_y \left(\frac{\partial x}{\partial y}\right)_z \left(\frac{\partial y}{\partial z}\right)_x + 1$$

and on rearranging we obtain:

$$\left(\frac{\partial x}{\partial y}\right)_z \left(\frac{\partial y}{\partial z}\right)_x \left(\frac{\partial z}{\partial x}\right)_y = -1$$

This cyclic relationship can also be written as its reciprocal

$$\left(\frac{\partial y}{\partial x}\right)_z \left(\frac{\partial z}{\partial y}\right)_x \left(\frac{\partial x}{\partial z}\right)_y = \left(\frac{\partial x}{\partial z}\right)_y \left(\frac{\partial y}{\partial x}\right)_z \left(\frac{\partial z}{\partial y}\right)_x = -1$$

These equations are most easily remembered by writing x, y, and z in some order, say alphabetical, and placing below them x, y, and z so that none of them occurs twice in the same column. There are two possible arrangements

$$\begin{matrix} x & y & z \\ y & z & x \end{matrix} \qquad\qquad \begin{matrix} x & y & z \\ z & x & y \end{matrix}$$

These arrangements then give the sequence of partial derivatives in the cyclic relationship.

$$\left(\frac{\partial x}{\partial y}\right)_z \cdot \left(\frac{\partial y}{\partial z}\right)_x \cdot \left(\frac{\partial z}{\partial x}\right)_y = -1 = \left(\frac{\partial x}{\partial z}\right)_y \cdot \left(\frac{\partial y}{\partial x}\right)_z \cdot \left(\frac{\partial z}{\partial y}\right)_x$$

The relationship can be used for example to derive an equation relating the Joule-Thomson coefficient (μ_{JT}) to the heat capacity at constant pressure (C_p) and the variation of enthalpy with pressure at constant temperature, $(\partial H/\partial P)_T$

$$\mu_{JT} = \frac{(\partial H/\partial P)_T}{C_p}.$$

To obtain the cyclic relationship of the partial differentials H, T and P, we arrange H, T, and P as described above for x, y, and z.

$$\begin{matrix} H & T & P \\ T & P & H \end{matrix}$$

i.e. $$\left(\frac{\partial H}{\partial T}\right)_P \cdot \left(\frac{\partial T}{\partial P}\right)_H \cdot \left(\frac{\partial P}{\partial H}\right)_T = -1$$

Now by definition

$$C_p = \left(\frac{\partial H}{\partial T}\right)_P \quad \text{and} \quad \mu_{JT} = \left(\frac{\partial T}{\partial P}\right)_H$$

therefore $$C_p\,\mu_{JT} \cdot \left(\frac{\partial P}{\partial H}\right)_T = -1$$

i.e. $$\mu_{JT} = \frac{-(\partial H/\partial P)_T}{C_p}$$

2.13 Maxima and Minima

Let us consider the function $y = x^3 - 4x^2 + 4x$. The curve of this function is plotted in Fig. 2.4 as a solid line and that of $dy/dx = 3x^2 - 8x + 4$ as a broken line.

It is obvious from the graph of $y = x^3 - 4x^2 + 4x$ that this function has two 'turning points' A and B, the former being described as a *maximum* and the latter as a *minimum*.

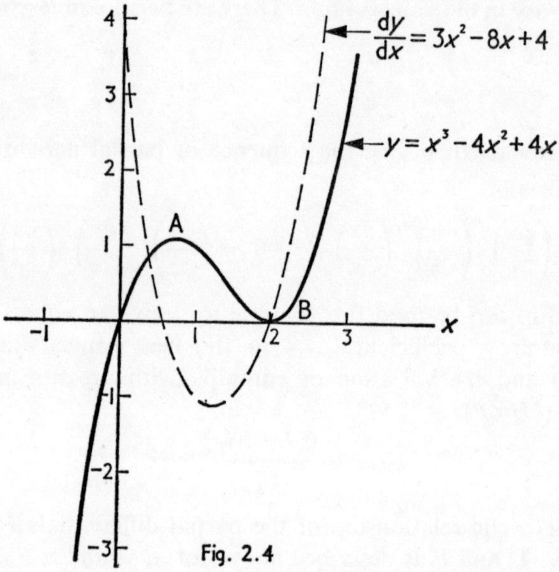

Fig. 2.4

[N.B. In this case these are not maximum or minimum values of the function. y tends to ∞ as x tends to ∞ and to $-\infty$ as x tends to $-\infty$. It is usually quite simple to show by plotting the curve of the function whether the turning point represents a maximum or minimum value of the function.]

A *turning point* is defined as a point at which the slope of the tangent to the curve changes sign. This is easily seen from the graph of dy/dx in Fig. 2.4. For values of x less than that corresponding to A, dy/dx has a value >0 and the tangent to the curve has a positive slope. At A, $dy/dx = 0$, i.e. the tangent is parallel to the x-axis. Between A and B, $dy/dx < 0$, i.e. the tangent has a negative slope. At B, $dy/dx = 0$, and for values of x greater than that corresponding to B, dy/dx is again positive.

To find the values of x corresponding to turning points, we require to find the values of x at which $dy/dx = 0$

i.e. $$dy/dx = 3x^2-8x+4 = 0$$

$$(3x-2)(x-2) = 0$$

$$x = \tfrac{2}{3} \quad \text{or} \quad 2$$

i.e. turning points occur at $x = \frac{2}{3}$ and $x = 2$.

To find if these correspond to maxima or minima we can either plot the curve of $y = x^3-4x^2+4x$ or apply the *second derivative test*. Consider the graph of dy/dx, where at the point A this function is decreasing i.e. the tangent to the curve at the point corresponding to the maximum turning point in $y = x^3-4x^2+4x$ would have a negative gradient or

$$\frac{d}{dx}(dy/dx) < 0$$

at a value of x corresponding to a maximum.

Note that $\frac{d}{dx}(dy/dx)$ is usually written d^2y/dx^2.

If $\quad dy/dx = 3x^2-8x+4$

$$d^2y/dx^2 = 6x-8$$

at $x = \frac{2}{3}$ $\quad d^2y/dx^2 = 6(\frac{2}{3})-8 = -4$

i.e. the turning point at $x = \frac{2}{3}$ is a maximum. Conversely at the minimum turning point, B, the slope of the tangent to dy/dx is positive

i.e. $\quad d^2y/dx^2 > 0$

at $x = 2$ $\quad d^2y/dx^2 = 6(2)-8 = +4$

i.e. the turning point at $x = 2$ is a minimum.

[N.B. It is possible to have a point of inflection which has a tangent parallel to the x-axis ($dy/dx = 0$). However, in this case d^2y/dx^2 would also be 0. See Fig. 2.5 where $y = x^3$ and $dy/dx = 3x^2$.]

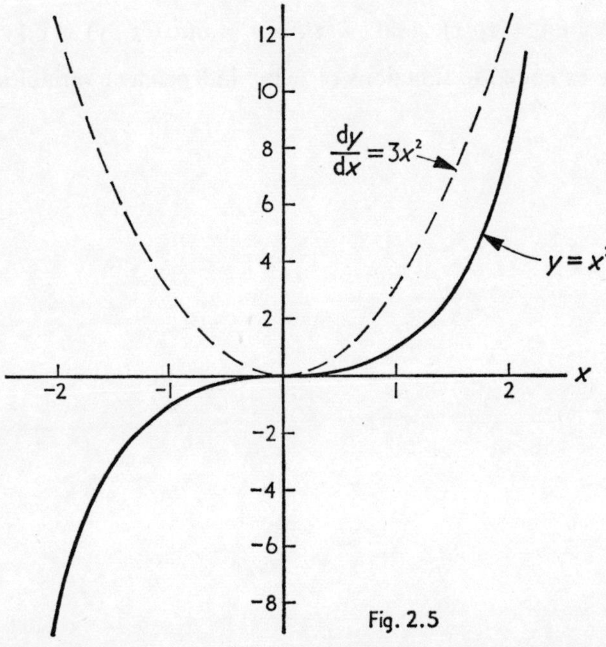

Fig. 2.5

2.14 Maxima and minima of functions of more than one variable

The conditions that a function $z = f(x, y)$ should have a turning point are

$$\partial z/\partial x = 0 \quad \text{and} \quad \partial z/\partial y = 0.$$

Such a point can be a maximum, a minimum or a 'saddle point' i.e. a point on a surface where one section through the surface at the point is a minimum and an orthogonal section is a maximum.

The tests which are applied to show whether a point on a surface represents a maximum or a minimum or neither are difficult to apply.

For a maximum at a point

$$\partial z/\partial x = 0, \quad \partial z/\partial y = 0, \quad \partial^2 z/\partial x^2 < 0, \quad \partial^2 z/\partial y^2 < 0$$

and $$(\partial^2 z/\partial x^2)(\partial^2 z/\partial y^2) > (\partial^2 z/\partial x \partial y)^2.$$

For a minimum

$$\partial z/\partial x = 0, \quad \partial z/\partial y = 0, \quad \partial^2 z/\partial x^2 > 0, \quad \partial^2 z/\partial y^2 > 0$$

and $$(\partial^2 z/\partial x^2)(\partial^2 z/\partial y^2) > (\partial^2 z/\partial x \partial y)^2.$$

If $(\partial^2 z/\partial x^2)(\partial^2 z/\partial y^2) < (\partial^2 z/\partial x \partial y)^2$ the point is neither a maximum nor a minimum.

$$[\partial^2 z/\partial x^2 \equiv \partial/\partial x(\partial z/\partial x) \quad \text{and} \quad \partial^2 z/\partial x \partial y = \partial/\partial x(\partial z/\partial y) \equiv \partial/\partial y(\partial z/\partial x).]$$

This can be extended to functions of many independent variables.

3 *Integral Calculus*

We have shown how to find the *differential coefficient* of a continuous function of x, i.e. the rate of change of the function. Often we only know the rate of change of the function and are required to calculate the function. This process is known as *integration*.

There are two definitions of an *integral*. One, as indicated above, is the inverse of a differential coefficient and the other is the limit of the sum of a series. The former definition enables us to evaluate the integral, while many of the applications depend on the latter.

3.1 The indefinite integral

The indefinite integral of a function of x, $f(x)$, with respect to x is written

$$\int f(x)\,dx$$

You will recall that if $y = x^2$, $dy/dx = 2x$.

Since integration is the inverse of differentiation, you would expect that

$$\int 2x\,dx = x^2$$

However, if for example, $y = x^2+10$ or indeed $y = x^2+a$, where a is not a function of x, in both cases

$$dy/dx = 2x$$

therefore $$\int 2x\,dx = x^2+c \quad \text{where } c \text{ is a constant.}$$

Often, this constant of integration is omitted, but its existence must be borne in mind.

3.2 Evaluation of some standard integrals

In order to evaluate even moderately complicated expressions, it is necessary to reduce the expression to integrals derived from a knowledge of differential coefficients. These integrals are commonly known as *standard integrals* and a list is given in Appendix C.

We will now show how some of these are derived.

Example. Evaluate $\int x^n \, dx$.

If
$$y = x^{n+1}$$
$$dy/dx = (n+1)x^n$$

therefore if
$$y = \frac{x^{n+1}}{n+1}$$
$$dy/dx = \frac{(n+1)}{(n+1)} x^n = x^n$$

therefore
$$\int x^n \, dx = y = \frac{x^{n+1}}{n+1} \text{ (the constant is omitted).}$$

This is true for all values of n except $n = -1$.

In this case, we know that $d/dx \ln x = 1/x = x^{-1}$

therefore
$$\int x^{-1} \, dx = \ln x$$

Thus we have derived the first two of our standard integrals, namely,

$$\int x^n \, dx = \frac{x^{n+1}}{n+1} \quad \text{for all values of } n \text{ except } n = -1,$$

and
$$\int x^{-1} \, dx = \ln x$$

Example. Evaluate the following integrals

$$\int x^6 \, dx = \frac{x^7}{7}$$

$$\int \frac{1}{x^6} \, dx = \int x^{-6} \, dx = -\tfrac{1}{5} x^{-5} = \frac{-x^{-5}}{5}$$

$$\int \sqrt[3]{x} \, dx = \int x^{\frac{1}{3}} \, dx = \frac{1}{4/3} x^{\frac{4}{3}} = \tfrac{3}{4} \sqrt[3]{x^4}$$

$$\int \frac{1}{\sqrt{x}} \, dx = \int x^{-\frac{1}{2}} \, dx = \frac{1}{1/2} x^{\frac{1}{2}} = 2\sqrt{x}.$$

3.3 Integration of a polynomial

We will now evaluate $\int (ax^3 + bx^2 + cx + e) \, dx$.

If
$$y = ax^4 + bx^3 + cx^2 + ex$$
$$dy/dx = 4ax^3 + 3bx^2 + 2cx + e$$

It follows that

$$\int (4ax^3+3bx^2+2cx+e)\,\mathrm{d}x = ax^4+bx^3+cx^2+ex$$

i.e.

$$\int (4ax^3+3bx^2+2cx+e)\,\mathrm{d}x = \int 4ax^3\,\mathrm{d}x+\int 3bx^2\,\mathrm{d}x+\int 2cx\,\mathrm{d}x+\int e\,\mathrm{d}x$$

therefore

$$\int (ax^3+bx^2+cx+e)\,\mathrm{d}x = \frac{ax^4}{4}+\frac{bx^3}{3}+\frac{cx^2}{2}+ex$$

i.e. the integral of any polynomial in x can immediately be integrated by summing the integrals of the individual terms.

Example. Evaluate the following integrals

$$\int (3x^6+8x^3-5)\,\mathrm{d}x = \frac{3x^7}{7}+\frac{8x^4}{4}-5x$$
$$= \frac{3x^7}{7}+2x^4-5x$$

$$\int (x-2)^3\,\mathrm{d}x = \int (x-2)(x^2-4x+4)\,\mathrm{d}x$$
$$= \int (x^3-6x^2+12x-8)\,\mathrm{d}x$$
$$= \frac{x^4}{4}-2x^3+6x^2-8x$$

$$\int \frac{(x-2)^3}{x^2}\,\mathrm{d}x = \int \frac{x^3-6x^2+12x-8}{x^2}\,\mathrm{d}x$$
$$= \int (x-6+12x^{-1}-8x^{-2})\,\mathrm{d}x$$
$$= \frac{x^2}{2}-6x+12\ln x+\frac{8}{x}$$

3.4 Change of variable

Normally, integration will involve the reduction of the integral to a combination of standard forms. We will now consider methods of carrying this out. The first method we will discuss is *change of variable.*

Example. Evaluate

$$\int \sin^4 x \cos x\,\mathrm{d}x$$

Let us consider $\quad u = \sin x$

Differentiating, $\quad \mathrm{d}u/\mathrm{d}x = \cos x$

Substituting in the original integral $u = \sin x$

and $$du/dx = \cos x$$

We have now to evaluate

$$\int (u^4 \, du/dx) \, dx = \int u^4 \, du$$

$$= \frac{u^5}{5} = \frac{\sin^5 x}{5}$$

Example. Evaluate

$$\int \frac{x^5}{a^6+x^6} \, dx$$

Let us consider $$u = a^6+x^6$$

$$du/dx = 6x^5$$

On substitution the integral becomes

$$\int (1/u)(\tfrac{1}{6}du/dx) \, dx = \tfrac{1}{6} \int (1/u) \, du$$

$$= \tfrac{1}{6} \ln u$$

$$= \tfrac{1}{6} \ln (a^6+x^6)$$

3.5 Integration by parts

Another important method of integration is *integration by parts*. This is the converse of finding the differential coefficient of the product of two functions of x.

i.e. $$\frac{d}{dx}(uv) = v\frac{du}{dx} + u\frac{dv}{dx}$$

integrating $$uv = \int v \frac{du}{dx} \, dx + \int u \frac{dv}{dx} \, dx$$

Rearranging $$\int u \frac{dv}{dx} \, dx = uv - \int v \frac{du}{dx} \, dx \qquad \text{...(3A)}$$

The following examples illustrate the use of this method.

Example. Evaluate

$$\int x^2 \ln x \, dx$$

Take $u = \ln x$ whence

$$du/dx = 1/x$$

and $dv/dx = x^2$ whence

$$v = \int x^2\,dx = x^3/3$$

Substituting in equation (3A)

$$\int x^2 \ln x\,dx = \frac{x^3 \ln x}{3} - \int \frac{x^3}{3}\left(\frac{1}{x}\right) dx$$

$$= \frac{x^3 \ln x}{3} - \frac{1}{3}\int x^2\,dx = \frac{x^3 \ln x}{3} - \frac{x^3}{9}$$

Example. Evaluate

$$\int x \sin x\,dx$$

Take $u = \sin x$ whence

$$du/dx = \cos x$$

and $dv/dx = x$ whence $\quad v = \dfrac{x^2}{2}$

Substituting in equation (3A)

$$\int x \sin x\,dx = \frac{x^2 \sin x}{2} - \int \frac{x^2 \cos x}{2}\,dx$$

We see that the right hand integral is not simple and hence an incorrect substitution has been made.

Take $u = x$ whence $\quad du/dx = 1$

and $dv/dx = \sin x$ whence

$$v = -\cos x$$

Substituting in equation (3A)

$$\int x \sin x\,dx = -x \cos x - \int -\cos x\,dx$$

$$= -x \cos x + \sin x$$

3.6 Successive reduction

A further extension of integration by parts is *successive reduction.*

Example. Evaluate

$$\int x^n \cos ax\,dx$$

Take $u = x^n$ whence

$$du/dx = nx^{n-1}$$

and $dy/dx = \cos ax$ whence

$$v = 1/a \sin ax$$

Substituting in equation (3A)

$$\int x^n \cos ax \, dx = \frac{x^n}{a} \sin ax - \int \frac{n}{a} x^{n-1} \sin ax \, dx$$

We now evaluate $\frac{n}{a} \int x^{n-1} \sin ax \, dx$ by the same method.

Take $u = x^{n-1}$ whence

$$du/dx = (n-1)x^{n-2}$$

and $dv/dx = \sin ax$ whence

$$v = -1/a \cos ax$$

therefore

$$\int x^{n-1} \sin ax \, dx = -\frac{x^{n-1}}{a} \cos ax + \frac{n-1}{a} \int x^{n-2} \cos ax \, dx$$

therefore

$$\int x^n \cos ax \, dx = \frac{x^n}{a} \sin ax + \frac{nx^{n-1}}{a^2} \cos ax - \frac{n(n-1)}{a^2} \int x^{n-2} \cos ax \, dx$$

The process is repeated until the integral on the right hand side is either

$$\int \cos ax \, dx \quad \text{or} \quad \int \sin ax \, dx$$

3.7 Definite integrals

The integrals we have discussed so far are *indefinite* integrals and we will now turn our attention to *definite* integrals, discussing, firstly, integration as a sum of a series and then as a means of calculating the area under a curve.

A definite integral is written

$$\int_a^b f(x) \, dx$$

a and b are called the *limits of integration* and the expression means the integral of $f(x)$ with respect to x between the limits $x = a$ and $x = b$. The value is calculated by integrating $f(x)$ in the usual way. Let the integral be $F(x)$. We subtract the value of $F(x)$ when $x = a$, $F(a)$, from the value of $F(x)$ when $x = b$, $F(b)$. The stage when the integration has been carried out but before substitution of $x = a$ and $x = b$ is often written as $[F(x)]_a^b$

i.e. $$\int_a^b f(x) \, dx = [F(x)]_a^b = F(b) - F(a).$$

The definition of a definite integral as the limit of a sum of a series is

$$\int_a^b f(x)\,dx = \lim_{h\to 0} \sum_{x=a}^{b} f(x)\,\delta x$$

The curve $y = f(x)$ is divided into lengths h along the x-axis between $x = a$ and $x = b$. The integral equals the sum when the length, h, tends to zero.

The integral can also be shown to be equal to the area under the curve $y = f(x)$ between $x = a$ and $x = b$ as shown in Fig. 3.1.

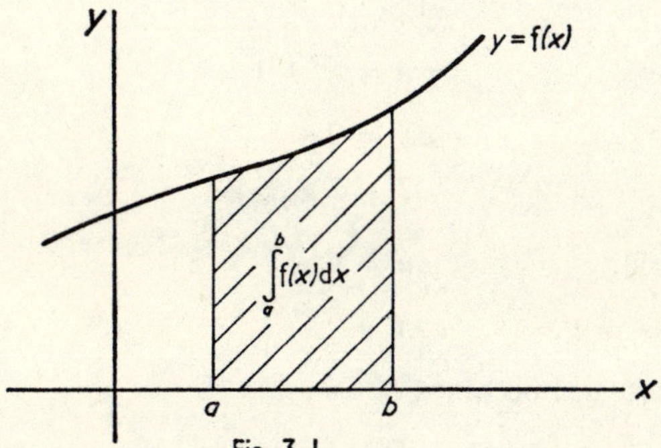

Fig. 3.1

The graph in Fig. 3.2 shows the variation of current with time at an electrode in an electrochemical cell under certain conditions.

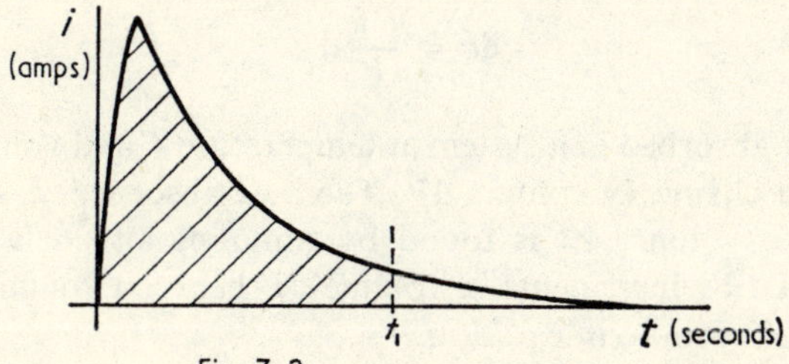

Fig. 3.2

In this case, $i = f(t)$ and the area under the curve,

$$q = \int_0^{t_1} i\,dt$$

corresponds to the total charge (in coulombs) passing from the start of the electrolysis until time t_1.

Example. Calculate the area bounded by $y = x^2$ and the x-axis between $x = 6$ and $x = 9$.

$$\text{Area} = \int_6^9 x^2\,\mathrm{d}x = \left[\frac{x^3}{3}\right]_6^9 = \frac{9^3}{3} - \frac{6^3}{3} = 243 - 72 = 171$$

If the integral is evaluated by change of variable then the limits must be changed.

Example. Evaluate

$$\int_0^4 \frac{x}{\sqrt{(x^2+9)}}\,\mathrm{d}x$$

Put $u = x^2+9$

then $\mathrm{d}u/\mathrm{d}x = 2x$

i.e. $x = \frac{1}{2}\,\mathrm{d}u/\mathrm{d}x$

When $x = 4$, $u = 4^2+9 = 25$

and when $x = 0$, $u = 9$

Thus the integral becomes

$$\int_9^{25} \frac{1}{\sqrt{u}}\left(\tfrac{1}{2}\mathrm{d}u/\mathrm{d}x\right)\mathrm{d}x = \tfrac{1}{2}\int_9^{25} \frac{1}{\sqrt{u}}\,\mathrm{d}u = \left[\tfrac{1}{2}.2\sqrt{u}\right]_9^{25} = 5-3 = 2$$

We will now consider some further examples of integrals used in chemistry.

Isothermal change in an ideal gas

$$\mathrm{d}q = \frac{RT}{V}\,\mathrm{d}V$$

$\mathrm{d}q$ is the heat absorbed in a system at temperature T and volume V when it experiences a change in volume $\mathrm{d}V$. The heat absorbed, q, in going from volume V_1 to volume V_2 is found by summing all small increments in heat $\mathrm{d}q$ caused by increments in volume $\mathrm{d}V$ between volumes V_1 and V_2 i.e. this is equivalent to integration.

$$q = \int_{V_1}^{V_2} \frac{RT}{V} = RT\int_{V_1}^{V_2} \frac{1}{V}\,\mathrm{d}V$$

$$= RT\,[\ln V]_{V_1}^{V_2}$$

$$= RT\,(\ln V_2 - \ln V_1)$$

$$= RT\ln\frac{V_2}{V_1}$$

Temperature dependence of the equilibrium constant

For changes in temperature dT

$$\frac{\mathrm{d}\ln K_p}{\mathrm{d}T} = \frac{\Delta H^0}{RT^2}$$

therefore $$\mathrm{d}\ln K_p = \frac{\Delta H^0}{R}\left(\frac{1}{T^2}\,\mathrm{d}T\right)$$

Let K_p' and K_p'' be the values of K_p at temperatures T_1 and T_2 respectively. To find the relationship between the equilibrium constants at T_1 and T_2 we integrate both sides of the equation.

i.e. $$\int_{K_p'}^{K_p''} \mathrm{d}\,(\ln K_p) = \frac{\Delta H^0}{R}\int_{T_1}^{T_2} \frac{1}{T^2}\,\mathrm{d}T$$

$$[\ln K_p]_{K_p'}^{K_p''} = \frac{\Delta H^0}{R}\left[-\frac{1}{T}\right]_{T_1}^{T_2}$$

therefore $$\ln K_p'' - \ln K_p' = -\frac{\Delta H^0}{R}\left(\frac{1}{T_2}-\frac{1}{T_1}\right)$$

i.e. $$\ln K_p''/K_p' = -\frac{\Delta H^0}{R}\left(\frac{1}{T_2}-\frac{1}{T_1}\right)$$

Example. Calculate N such that the electron whose wave function $\Psi = N\sin(\pi rx/a)$, where $r = 1, 2, 3\ldots$, has unit probability of being found at a point between $x = 0$ and $x = a$.

We require to evaluate N such that

$$\int_0^a N^2 \sin^2(\pi rx/a)\,\mathrm{d}x = 1$$

We will simplify the integral by letting $\pi r/a = b$

i.e. $$N^2 \int_0^a \sin^2 bx\,\mathrm{d}x = 1$$

To evaluate this integral we require the equation

$$\sin^2 bx = \tfrac{1}{2}(1-\cos 2bx)$$

The integral can now be evaluated.

$$\tfrac{1}{2}N^2 \int_0^a (1-\cos 2bx)\,\mathrm{d}x = 1$$

$$\tfrac{1}{2}N^2\left[x-\frac{1}{2b}\sin 2bx\right]_0^a = 1$$

Replacing b by $\pi r/a$

$$\tfrac{1}{2}N^2\left[x-\frac{a}{2\pi r}\sin\frac{2\pi rx}{a}\right]_0^a = 1$$

i.e.
$$\tfrac{1}{2}N^2\left(a-\frac{a}{2\pi r}\sin 2\pi r-\frac{a}{2\pi r}\sin 0\right) = 1$$

Now $\sin 0 = 0$ and $\sin 2\pi r = 0$ if r is an integer

i.e.
$$\tfrac{1}{2}N^2 a = 1$$
$$N = \pm\sqrt{(2/a)}$$

3.8 Multiple integration

Occasionally we require to evaluate multiple integrals. In general these are written

$$\iiint\ldots\int f(x_1x_2x_3\ldots x_n)\,dx_1\,dx_2\,dx_3\,\ldots\,dx_n$$

This involves successive integration of the function with respect to $x_1x_2x_3\ldots x_n$ in turn. If limits are different for different variables $x_1x_2x_3\ldots x_n$ the limit on the innermost integral applies to x_1 and the next to x_2 etc.

Example. Evaluate

$$\int_0^{2\pi}\int_0^{\pi}\sin\theta\,d\theta\,d\phi$$

First we have to integrate with respect to θ using the limits 0 to π.

$$\int_0^{2\pi}\int_0^{\pi}\sin\theta\,d\theta\,d\phi = \int_0^{2\pi}[-\cos\theta]_0^{\pi}\,d\phi$$
$$= \int_0^{2\pi}(-\cos\pi+\cos 0)\,d\phi$$

Now $\cos\pi = -1$ and $\cos 0 = 1$

therefore
$$\int_0^{2\pi}\int_0^{\pi}\sin\theta\,d\theta\,d\phi = \int_0^{2\pi}(1+1)\,d\phi$$
$$= [2\phi]_0^{2\pi}$$
$$= 4\pi$$

Example. Calculate the mean value of $1/r$, $\overline{(1/r)}$, for an electron in the $1s$ orbital of a one electron atom using the equation

$$\overline{1/r} = \int_0^{2\pi}\int_0^{\pi}\int_0^{\infty} 1/\pi(z/a)^3 e^{-2zr/a_0}\, r\sin\theta\,dr\,d\theta\,d\phi$$

First we integrate with respect to r.

$\int re^{-2zr/a_0}\,\mathrm{d}r$ may be solved using integration by parts.

$$\int u\frac{\mathrm{d}v}{\mathrm{d}r}\,\mathrm{d}r = uv - \int v\frac{\mathrm{d}u}{\mathrm{d}r}\,\mathrm{d}r$$

Let $u = r$ then $\qquad \mathrm{d}u/\mathrm{d}r = 1$

and $\mathrm{d}v/\mathrm{d}r = e^{-2zr/a_0}$

$$v = \int e^{-2zr/a_0}\,\mathrm{d}r = \frac{-a_0}{2z}\,e^{-2zr/a_0}$$

Thus
$$\int re^{-2zr/a_0}\,\mathrm{d}r = \frac{-a_0 r}{2z}\,e^{-2zr/a_0} - \int -\frac{a_0}{2z}\,e^{-2zr/a_0}\,\mathrm{d}r$$
$$= \frac{-a_0 r}{2z}\,e^{-2zr/a_0} + \frac{a_0}{2z}\left(-\frac{a_0}{2z}\,e^{-2zr/a_0}\right)$$
$$= \frac{-a_0}{2z}\,e^{-2zr/a_0}\left(r + \frac{a_0}{2z}\right)$$

therefore

$$\overline{1/r} = \int_0^{2\pi}\int_0^{\pi}\left(\frac{z}{a_0}\right)^3\frac{\sin\theta}{\pi}\left[-\frac{a_0}{2z}\,e^{-2zr/a_0}\left(r + \frac{a_0}{2z}\right)\right]_0^{\infty}\mathrm{d}\theta\,\mathrm{d}\phi$$
$$= \int_0^{2\pi}\int_0^{\pi}\left(\frac{z}{a_0}\right)^3\frac{\sin\theta}{\pi}\left(0 + \left(\frac{a_0}{2z}\right)^2\right)\mathrm{d}\theta\,\mathrm{d}\phi \qquad \text{since } e^{-2zr/a_0} = 0 \text{ when } r = \infty$$

therefore

$$\overline{1/r} = \int_0^{2\pi}\int_0^{\pi}\frac{z}{4a_0\pi}\sin\theta\,\mathrm{d}\theta\,\mathrm{d}\phi$$
$$= \int_0^{2\pi}\frac{z}{4a_0\pi}\,[-\cos\theta]_0^{\pi}\,\mathrm{d}\phi$$

Now, $\cos\pi = -1$ and $\cos 0 = 1$

therefore

$$\overline{1/r} = \int_0^{2\pi}\frac{z}{4a_0\pi}.2\,\mathrm{d}\phi$$
$$= \frac{z}{2a_0\pi}\,[\phi]_0^{2\pi}$$
$$= \frac{z}{a_0}$$

4 *Differential Equations*

4.1 Introduction

In many areas of chemistry one is concerned with rates of change: thus for example in the study of reaction kinetics one studies rates of change of concentration with time, while in the investigation of chemical bonding one may wish to calculate the rate of change of electron density probability as a function of distance. Any equation containing a rate (i.e. a *derivative*) is called a *differential equation.*

There are a number of terms used to describe such equations. The *order* of a differential equation is the order of the highest derivative in it; the *degree* is the index or exponent of the highest order derivative when the equation is written as a simple polynomial. Thus for example

$$y'' = 3y^2 - x^3$$

is a *second order first degree* equation, while

$$(y')^2 = y - y'$$

is a *first order second degree* equation.

A *linear* equation has the form

$$ay + by' + cy'' + \dots = k$$

where a, b, $c \dots k$ are constants or functions only of x, the independent variable. Thus the equations

$$3xy' = y \quad \text{and} \quad y'' = 3y' - x$$

are linear, while equations $3x(y')^2 = y$ and $y'' = 3yy' - x$ are not.

A solution to a differential equation is any function of the independent variable which satisfies the equation. There are an infinite number of solutions for any one equation since a linear equation of order n is found to have a solution containing n arbitrary constants. Such a solution is called the *general solution.* More important in the practical application of differential equations are *particular solutions* where the values of the constants are assigned so that the solution corresponds to known physical conditions.

Consider the simple first order kinetic rate law for the reaction

$$A \rightarrow \text{Products},$$

$$\frac{-\mathrm{d}[A]}{\mathrm{d}t} = k[A]$$

The general solution to this differential equation (as may be verified by substitution) is

$$[A] = C \exp(-kt)$$

where C is a constant.

A particular solution is found when we impose the condition that $[A] = [A]_0$ at $t = 0$. Then $C = [A]_0$, and we have

$$[A] = [A]_0 \exp(-kt)$$

or, in its more familiar form,

$$\ln [A] = -kt + \ln [A]_0$$

The conditions which must be satisfied by a particular solution are called *boundary conditions* and it is usually possible to derive the appropriate particular solution from the general solution by evaluating the constants from the boundary conditions, as in the example above. There is unfortunately no general rule for obtaining the general solution of a differential equation: in fact it is often impossible to find such a solution at all. We will consider only a number of special cases which are of interest in chemistry and where solutions may be readily found.

4.2 Separable variables

In this, the simplest class of differential equation, the equation may be written with only terms in y on one side and terms in x on the other. An example of such an equation is the reaction kinetics example studied above. Note that exactly the same equation is encountered when one considers, for example, the absorption of radiation as a function of the thickness of the absorber, or simple radioactive decay. Where the variables can be separated, the general solution is found simply by integrating each side of the equation.

Example. For hydrogen gas,

$$C_p = \mathrm{d}S/\mathrm{d}t = 29(T^{-1}) + 8{\cdot}3 \times 10^{-4}$$

where C_p is given in $\mathrm{JK^{-1}\,mol^{-1}}$. Find an expression for $S_{H_2(g)}$ as a function of T if $S^{298}_{H_2(g)}$, the entropy of the gas at 298 K, is $130{\cdot}5\ \mathrm{JK^{-1}\,mol^{-1}}$.

Separating the variables we have

$$dS = 29{\cdot}0\, dT/T + 8{\cdot}3\times 10^{-4}\, dT$$

Integrating each side produces the general solution

$$S = 29{\cdot}0 \ln T + 8{\cdot}3\times 10^{-4} T + C$$

where C is a constant.

If $S = 130{\cdot}5$ when $T = 298$ then $C = -35{\cdot}4$, and the particular solution, which is the required relationship, is

$$S = -35{\cdot}4 + 29{\cdot}0 \ln T + 8{\cdot}3\times 10^{-4} T$$

4.3 Linear equations of the first order

The general expression for such a differential equation is

$$ay' + by = c$$

where a, b and c are functions of x only or constants. By dividing throughout by a we have

$$y' + \frac{b}{a} y = \frac{c}{a}$$

or, as it is generally written,

$$y' + Py = Q$$

Again P and Q are functions of x only or constants.

One very useful method of finding particular solutions for this class of equation is to make use of the overall general solution

$$y \exp \int P\, dx = \int Q(\exp \int P\, dx)\, dx + C$$

Example. Solve $$xy' + 3y = 3x^2$$

Dividing throughout by x we have

$$y' + (3/x)y = 3x$$

Hence $$y\left(\exp \int \frac{3\, dx}{x}\right) = \int 3x \left(\exp \int \frac{3\, dx}{x}\right) dx + C$$

i.e. $$y(\exp (3 \ln x)) = \int 3x(\exp (3 \ln x))\, dx + C$$

therefore $$yx^3 = \int 3x^4\, dx + C$$

$$= \tfrac{3}{5} x^5 + C$$

and $$y = \tfrac{3}{5} x^2 + \frac{C}{x^3}$$

As in all such problems the solution may be verified by substitution in the original equation.

This class of equation has a number of chemical applications: for instance in the solution of kinetics problems where there are successive reactions, or in the parallel situation where a radioactive nuclide breaks down in a stepwise disintegration path.

Example. In the following reaction sequence

$$A \xrightarrow{k_1} B \xrightarrow{k_2} C,$$

what is the concentration of B at time t, if the initial concentration of A was $[A]_0$ and that of B and C was zero?

As above, $$-\frac{\mathrm{d}\,[A]}{\mathrm{d}t} = k_1[A],$$

and thus $$[A] = [A]_0 \exp(-k_1 t).$$

Note that $[A]$ is the concentration of A *at time t.*

Now the rate of removal of A must be identical to the rate of formation of B; however B is also being converted to C at a rate given by $k_2[B]$,

therefore $$\frac{\mathrm{d}\,[B]}{\mathrm{d}t} = k_1[A] - k_2[B]$$

$$\frac{\mathrm{d}\,[B]}{\mathrm{d}t} + k_2[B] = k_1[A]$$

and $$\frac{\mathrm{d}\,[B]}{\mathrm{d}t} + k_2[B] = k_1[A]_0 \exp(-k_1 t)$$

$$y' \quad + P \; y \quad = \quad Q$$

Since this equation is linear first order in $[B]$ and t, it may be solved in the usual manner:

$$[B] \exp k_2 t = \int (k_1[A]_0 \exp(-k_1 t) \exp(k_2 t))\,\mathrm{d}t + z,$$

where z is a constant

$$= \int k_1[A]_0 \exp(k_2 - k_1)t\,\mathrm{d}t + z$$

$$= \frac{k_1[A]_0}{k_2 - k_1} \exp(k_2 - k_1)t + z$$

This is the general solution. Now at $t = 0$, $[B] = 0$

therefore $$0 = \frac{k_1[A]_0}{k_2-k_1} + z, \qquad \text{since } \exp 0 = 1,$$

and $$z = \frac{-k_1[A]_0}{k_2-k_1}$$

Hence $$[B] \exp k_2 t = \frac{k_1[A]_0}{k_2-k_1}(\exp(k_2-k_1)t-1)$$

and $$[B] = \frac{k_1[A]_0}{k_2-k_1}\left(\frac{\exp(k_2-k_1)t}{\exp k_2 t} - \frac{1}{\exp k_2 t}\right)$$

$$= \frac{k_1[A]_0}{k_2-k_1}(\exp(-k_1 t)-\exp(-k_2 t))$$

The variation of $[A]$, $[B]$ and $[C]$ is shown in Fig. 4.1.

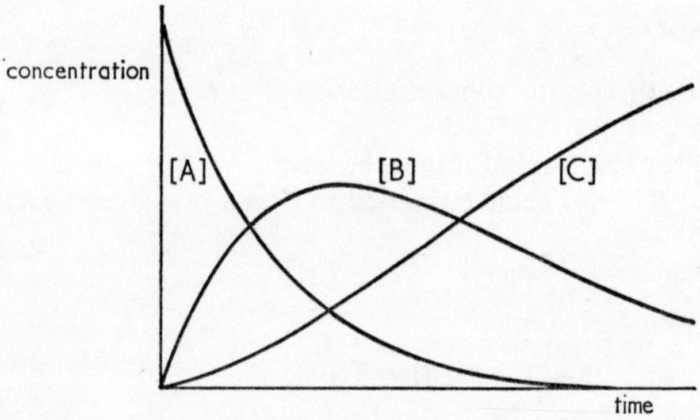

Fig. 4.1

Certain non-linear equations may be readily transformed into first order linear equations. Bernoulli's equation which has the form,

$$y' + Py = Qy^n$$

may be reduced to

$$z' + (1-n)Pz = (1-n)Q$$

where $z = y^{(1-n)}$.

4.4 Linear equations with constant coefficients

The solution of the equation

$$\frac{d^n y}{dx^n} + a_1 \frac{d^{n-1}y}{dx^{n-1}} + \ldots + a_{n-1}\frac{dy}{dx} + a_n y = f(x)$$

where $a_{1 \ldots n}$ are constants, is obtained most easily by the method of *Laplace transforms*. A good way of understanding the basis of this technique is to consider as an analogy the use of logarithms to simplify the process of multiplication. In this manipulation multiplication is effectively replaced by the easier process of addition. Thus to evaluate the product

$$p = 1{\cdot}23 \times 5{\cdot}01 \times 7{\cdot}94$$

we first *transform* the equation to

$$\begin{aligned} \log_{10}p &= \log_{10}1{\cdot}23 + \log_{10}5{\cdot}01 + \log_{10}7{\cdot}94 \\ &= 0{\cdot}09 + 0{\cdot}70 + 0{\cdot}90 \end{aligned}$$

so that $\log_{10}p$ can be found (by addition) to be 1·69. We now convert back to the original variable p by taking antilogarithms of each side, i.e. by using the *inverse transform*:

$$\log_{10}p = 1{\cdot}69$$

therefore $$p = 49{\cdot}0$$

The sequence may be summarised thus:

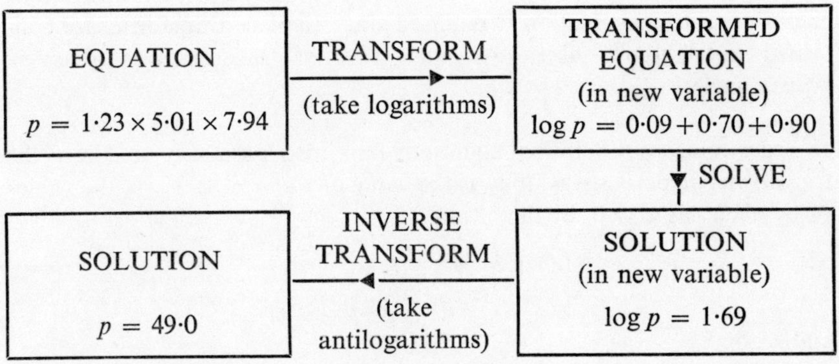

When one uses Laplace transforms, each term in a *differential* equation is substituted by an '*integral* transform'. This substitution produces an equation without derivatives in it, which can usually be easily solved, i.e. we replace the process of solving a differential equation by the easier process of solving an 'algebraic' equation. The latter, however, is not in the same variable as the original equation and the final step is therefore the conversion of the solution back to the original variable. Thus we have:

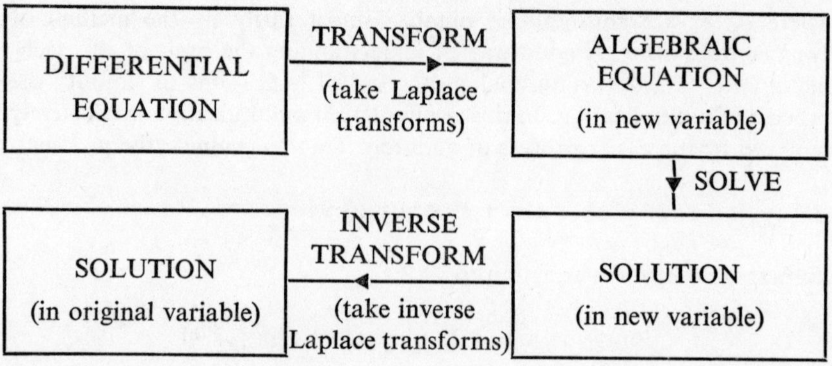

The Laplace transform L(y) of a function $y = \mathrm{f}(x)$ is defined as

$$\mathrm{L}(y) = \int_0^\infty \mathrm{f}(x) \exp(-px)\,\mathrm{d}x = \mathrm{F}(p)$$

Thus for example the Laplace transform of the function $\mathrm{f}(x) = 1$ is

$$\begin{aligned} \mathrm{L}(y) = \mathrm{F}(p) &= \int_0^\infty 1 \exp(-px)\,\mathrm{d}x \\ &= [-(1/p)\exp(-px)]_0^\infty \\ &= 1/p \quad \text{for} \quad p > 0 \end{aligned}$$

In a similar manner F(p) could be found for other much more complicated functions of x. In practice it is unnecessary to carry out such derivations each time a transformation is required since Laplace transforms are contained in standard tables. A short table of transforms is given in Appendix D.

In order to solve differential equations it is also necessary to know the transforms of derivatives. If y_0 is the value of y at $x = 0$, y'_0, y''_0 the values of y', y'' etc. at $x = 0$, by definition

$$\begin{aligned} \mathrm{L}(y') &= p(\mathrm{L}(y)) - y_0 \\ \mathrm{L}(y'') &= p^2(\mathrm{L}(y)) - py_0 - y'_0 \end{aligned}$$

and so on.

For simplicity of notation these are often written with

$$\begin{aligned} \mathrm{L}(y) &= Y \\ \mathrm{L}(y') &= Y' \quad \text{etc.} \end{aligned}$$

e.g.

$$\begin{aligned} Y' &= pY - y_0 \\ Y'' &= p^2 Y - py_0 - y'_0 \end{aligned}$$

Use of Laplace transforms for solving certain types of simple differential equation is very straightforward as can be seen from the following examples.

Example. Solve the equation $y''-6y'+9y = 2\exp(3x)$, if $y_0' = y_0 = 0$.

(i) Take Laplace transforms of each term:

$$(p^2Y-py_0-y_0')-6(pY-y_0)+9Y = 2\left(\frac{1}{p-3}\right)$$

The last term is found from tables. (See Appendix D.)

(ii) Substitute boundary conditions and solve for Y:

$$p^2Y-6pY+9Y = \frac{2}{p-3}$$

therefore $$(p^2-6p+9)Y = \frac{2}{p-3}$$

therefore $$Y = \frac{2}{(p-3)^3}$$

(iii) Find y, by obtaining inverse transform of Y:

$$Y = \frac{2}{(p-3)^3}$$

therefore $y = x^2\exp 3x$, from tables. (Appendix D.)

Again this solution may be verified by substitution.

It is not always quite so simple to find the inverse transform of Y. In most cases it is necessary to manipulate and combine the terms in the expression for Y to form factors whose transforms are known. (More advanced methods are available when such manipulations fail.)

Example. Transform:

$$Y = \frac{6}{p^2+4}$$

$$Y = \tfrac{6}{2}\left\{\frac{2}{p^2+2^2}\right\}$$

therefore $y = 3\sin 2x$.

Example. Transform:

$$Y = \frac{-p^2-8p+8}{p^4-4p^3+4p^2}$$

$$Y = \frac{-p^2-8p+8}{p^2(p-2)^2}$$

$$= \frac{-2(p-2)^2+p^2}{p^2(p-2)^2}$$

$$= \frac{1}{(p-2)^2}-\frac{2}{p^2}$$

$$= \frac{1}{(p-2)^2}-2\frac{1}{p^2}$$

therefore

$$y = x \exp 2x-2x$$

$$= x(\exp 2x-2)$$

One of the very important differential equations in chemistry describes the motion of a perfect harmonic oscillator. This equation is used in the discussion of bond vibrations and for one-dimensional wave equations.

Example. If the displacement of a particle from an origin is y and the restoring force $-ky$, the equation of motion of the particle is

$$m\frac{d^2y}{dt^2} = -ky$$

where m is the mass of the particle. Find the most general solution for y as a function of t.

The equation may be written

$$y''+\frac{k}{m}y = 0$$

Taking Laplace transforms we have

$$p^2Y-py_0-y_0'+\frac{k}{m}Y = 0\dagger$$

therefore

$$Y = \frac{y_0p+y_0'}{p^2+(k/m)}$$

$$= y_0\left\{\frac{p}{p^2+(k/m)}\right\}+y_0'\left\{\frac{1}{p^2+(k/m)}\right\}$$

---

†N.B.—y_0 is the position of the particle at $t = 0$ and y_0' is the velocity at $t = 0$.

Now taking inverse transforms,

$$y = y_0 \cos (k/m)^{\frac{1}{2}} + y_0' \sin (k/m)^{\frac{1}{2}}$$

4.5 Non-linear differential equations

It is often possible to reduce such equations to linear equations by substituting a new function for y. An example of this was given above where Bernoulli's equation was reduced to a first order linear equation.

Example. Reduce the equation

$$2y'' + 2(y')^2 = f(x)$$

to linear form.

Let $z = y'$, then $$z' = y''$$

thus $$2z' + 2z^2 = f(x)$$

i.e. $$z' + z^2 = \tfrac{1}{2}f(x)$$

4.6 Partial differential equations

While there is an infinite number of types of partial differential equation there is a small number of particular interest to chemists. We will not consider in detail the solutions of these equations here, although particular solutions have been found corresponding to many physical situations.

The use of integral transform methods is valuable in the solution of partial differential equations. Thus the Laplace transform of such an equation has one less independent variable than the equation: e.g. a partial differential equation in two independent variables may be reduced to a differential (not an algebraic) equation in one. This may then be solved in the normal fashion.

Many problems in electrochemistry are concerned with linear diffusion processes which are governed by an equation known as Fick's second law. One is interested in the variation in the concentration of a diffusing substance as a function of time t and position x. The relevant equation is

$$D \frac{\partial^2 c}{\partial x^2} = \frac{\partial c}{\partial t}$$

where c is concentration and D is the diffusion coefficient. If we take the Laplace transform in t, the variable x is then just a number.

Now $$L\left(\frac{\partial c}{\partial t}\right) = L(c') = pC - c_0$$

and $$L\left(\frac{\partial^2 c}{\partial x^2}\right) = \frac{\partial^2 C}{\partial x^2}$$

and thus the partial differential equation may be transformed to give

$$\frac{1}{D}(pC - c_0) = \frac{\partial^2 C}{\partial x^2}$$

This is now an ordinary differential equation in C (N.B. not c) and x which may be solved in the usual way.

We end this chapter by noting four of the most important partial differential equations in chemistry. Methods of finding particular solutions of these corresponding to specific physical situations will be found in most textbooks of physical chemistry, and these methods should be readily understood if the principles outlined above are applied.

(i) *Poisson's equation*

$$\nabla^2 g = f(x, y, z) \qquad \text{(see Section 7.13)}$$

g is a physical quantity which is a time invariant function of position (e.g. g may be the potential within a certain charge distribution).

(ii) *General diffusion equation*

$$\nabla^2 C = \frac{1}{D} \cdot \frac{\partial^2 C}{\partial t^2}$$

C is the concentration of a diffusing substance and D is a constant.

(iii) *General wave equation*

$$\nabla^2 X = \frac{1}{v^2} \cdot \frac{\partial^2 X}{\partial t^2}$$

X is the displacement from equilibrium of a vibrating component, and v is the velocity of propagation of the waves.

(iv) *Helmholtz equation*

$$\nabla^2 \psi + k\psi = 0$$

This is the time invariant wave equation used in connection with standing wave systems—e.g. in the time-independent Schrödinger equation.

5 *Coordinate Systems*

It is often necessary to fix the position of a point in either two or three dimensional space. To do this some form of reference framework must be selected and the position of the point specified relative to this framework.

5.1 The two dimensional case

If the point lies on a plane (i.e. a two dimensional case) we may draw two axes in the plane and call the intersection of these axes the origin (Fig. 5.1).

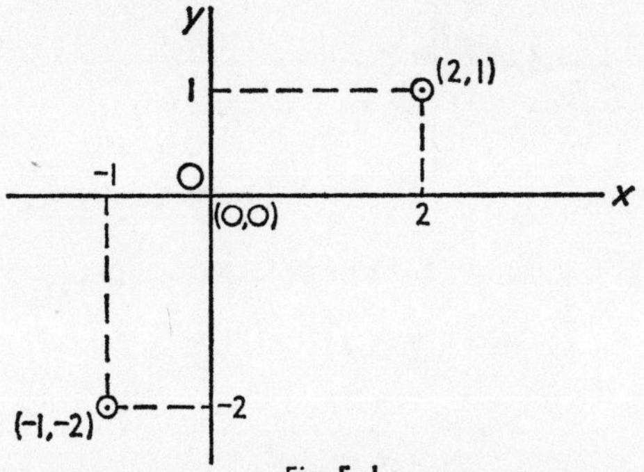

Fig. 5.1

For simplicity, we will make the axes mutually perpendicular, i.e. *orthogonal.* We can now specify any point in this plane by using two parameters x and y. These are known as *Cartesian coordinates.*

The distance d between any two points (x_1, y_1) and (x_2, y_2) is given by

$$d = \sqrt{[(x_1-x_2)^2+(y_1-y_2)^2]}$$

Further the distance r of any point (x, y) from the origin is then given by

$$r = \sqrt{(x^2+y^2)}$$

A second means of specifying the position of a point P is by using what are known as *polar coordinates.*

Let r be the distance of the point from the origin and θ the angle OP makes with a reference axis, i.e. the positive x-axis in Fig. 5.2.

D

The distance r will be given by

$$r = \sqrt{(x^2+y^2)}$$

and from trigonometry

$$\cos\theta = x/r$$

and $$\sin\theta = y/r$$

Thus $$x = r\cos\theta$$

and $$y = r\sin\theta$$

The coordinates (r, θ) are the polar coordinates of the point P.

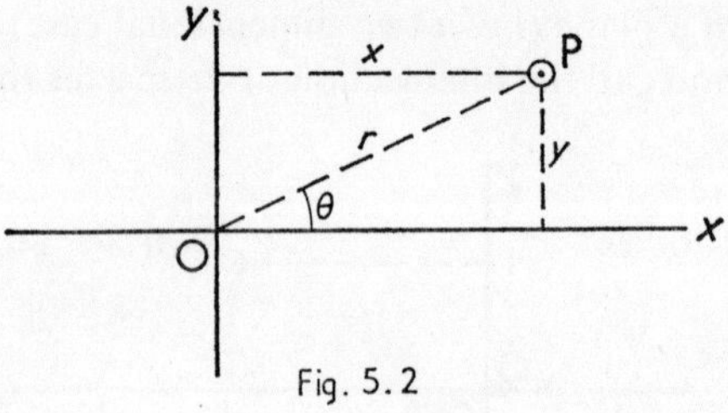

Fig. 5.2

Example. Express the point (1, 1) in polar coordinates:

$$r = \sqrt{(1^2+1^2)} = \sqrt{2}$$

$$\cos\theta = \frac{x}{r} = \frac{1}{\sqrt{2}}$$

$$\theta = \frac{\pi}{4} \text{ radians} = 45^\circ.$$

In polar coordinates the point is $(\sqrt{2}, \pi/4)$.

Note: θ is generally written in radians; π radians $= 180^\circ$.

5.2 The three dimensional case

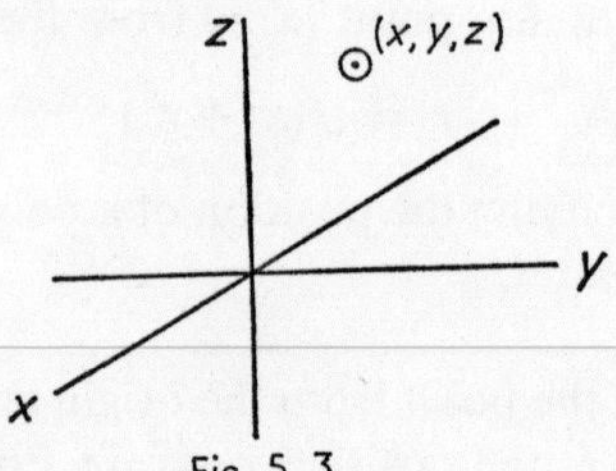

Fig. 5.3

To denote a point in three dimensional space we need a further parameter.

If we draw three orthogonal axes, Fig. 5.3, then the point may be specified by the Cartesian coordinates (x, y, z).

In addition to the Cartesian coordinate system two further systems, the *cylindrical polar* and *spherical polar* coordinate systems, are widely used.

(a) *Cylindrical polar coordinates* (r, θ, z)

The cylindrical coordinates are simply the polar coordinates in the xy plane with z as the third variable (Fig. 5.4).

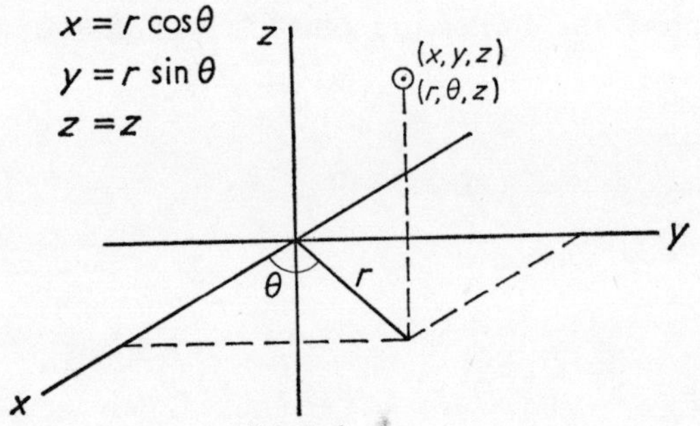

Fig. 5.4

Example. Express the Cartesian point (1, 1, 3) in cylindrical polar co-ordinates.

From the previous example we see that $r = \sqrt{2}$ and $\theta = \pi/4$. Therefore the cylindrical polar coordinates are $(\sqrt{2}, \pi/4, 3)$.

(b) *Spherical polar coordinates* (r, θ, ϕ)

These coordinates are made up of the distance from the origin r and two angles (see Fig. 5.5).

The point (x, y, z) in spherical polar coordinates is written (r, θ, ϕ), where

$$x = r \sin\theta \cos\phi$$

$$y = r \sin\theta \sin\phi$$

and

$$z = r \cos\theta$$

The use of spherical polar coordinates instead of Cartesian coordinates in certain cases considerably simplifies the solution of the Schrödinger equation.

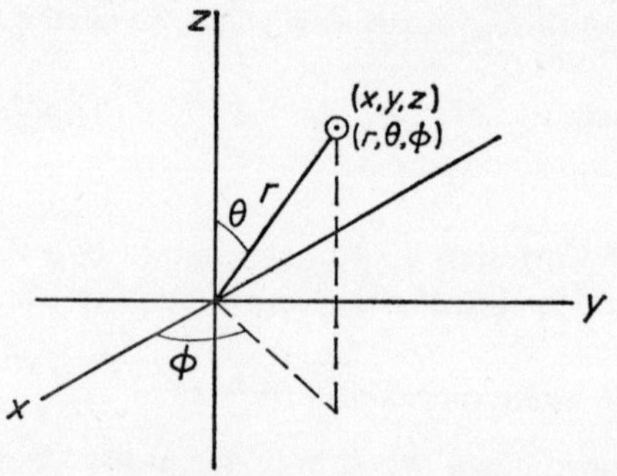

Fig. 5.5

Example. Express the Cartesian point (2, 1, 1) in spherical polar coordinates.

$$r = \sqrt{(2^2+1^2+1^2)}$$
$$= \sqrt{6} = 2{\cdot}449$$
$$\theta = \cos^{-1}(z/r)$$
$$= \cos^{-1}\left(\frac{1}{\sqrt{6}}\right) = \cos^{-1}(0{\cdot}4082)$$
$$= 65^\circ\ 54' = 1{\cdot}150 \text{ radians.}$$
$$\phi = \tan^{-1}(y/x)$$
$$= \tan^{-1}(0{\cdot}5)$$
$$= 26^\circ\ 34' = 0{\cdot}464 \text{ radians.}$$

The spherical polar coordinates of the point are (2·449, 1·150, 0·464).

5.3 Volume elements in the Cartesian, cylindrical and spherical coordinate systems

The volume element $\mathrm{d}V$ may be expressed in Cartesian coordinates as

$$\mathrm{d}V = \mathrm{d}x\,\mathrm{d}y\,\mathrm{d}z$$

in cylindrical polar coordinates as

$$\mathrm{d}V = r\,\mathrm{d}r\,\mathrm{d}\theta\,\mathrm{d}z$$

and in spherical polar coordinates as

$$\mathrm{d}V = r^2 \sin\theta\,\mathrm{d}r\,\mathrm{d}\theta\,\mathrm{d}\phi.$$

6 *Complex Numbers*

6.1 Introduction

Only positive numbers have real square roots, and if the square root of a negative number is required it must be written as an *imaginary number*. The imaginary number, i, is defined as

$$i = \sqrt{-1}, \quad \text{i.e. } i^2 = -1$$

Imaginary numbers, such as $3i$, $-ix$, etc., form a new set of numbers in addition to real numbers. A *complex number* is any number, real, imaginary or a combination of real and imaginary and can be written in general as

$$z = x + iy$$

Complex numbers obey the normal algebraic rules used for real numbers.

e.g.
$$(2+i)+(3-2i) = 5-i$$
$$(2+i)(3-2i) = 6-i-2i^2$$
$$= 8-i \qquad \text{since } i^2 = -1.$$

Complex numbers are of importance in simplifying, for example, the mathematics of quantum mechanics and electrical engineering.

Since complex numbers contain a real and an imaginary part, either of which may be zero, it is possible to plot them on a complex plane where the x-axis is the real axis and the y-axis the imaginary axis. Some examples are shown in Fig. 6.1.

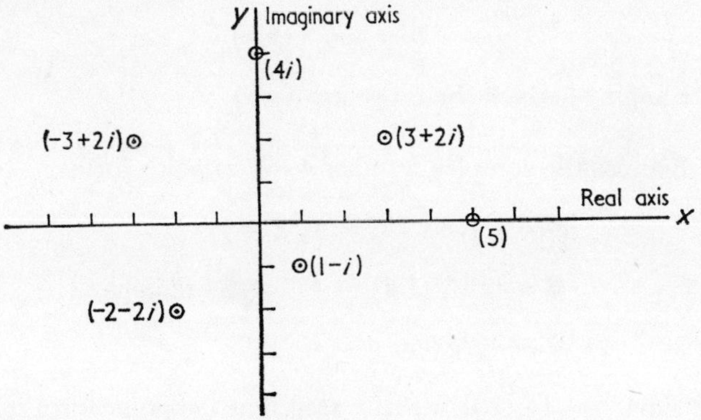

Fig. 6.1

6.2 Polar form of complex numbers

In a real plane the point (x, y) may also be denoted by (r, θ), (see Chapter 5). The analogous result for the complex plane will be

$$x = r\cos\theta$$

$$iy = ir\sin\theta$$

so that

$$z = x+iy = r(\cos\theta+i\sin\theta)$$

thus giving another means of expressing a complex number.

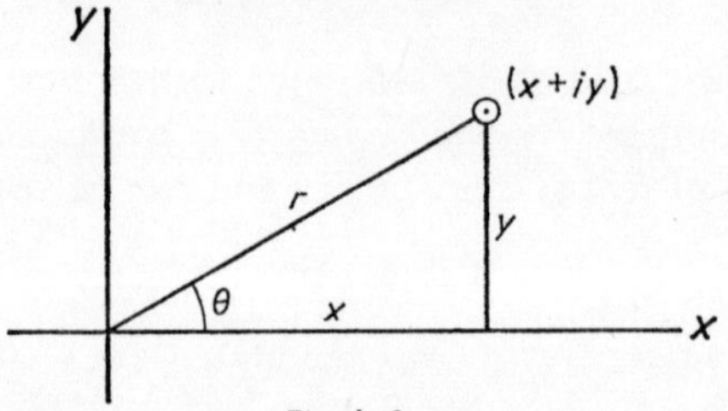

Fig. 6.2

In this polar form of the complex number, r is known as the *modulus* and θ as the *argument*. r is seen from Fig. 6.2 to be the distance from the origin to the point $(x+iy)$ and the length of this line will be $\sqrt{(x^2+y^2)}$,

$$r = \sqrt{(x^2+y^2)} = |z|$$

$|z|$ is known as the 'modulus of z' and is, by definition, always positive.

Also from Fig. 6.2 we see that

$$\tan\theta = y/x$$

$$\theta = \tan^{-1}(y/x)$$

i.e. θ is the angle of which the tangent is (y/x)

Example. Express the complex number $4+4i$ in polar form.

$$|z| = \sqrt{(16+16)} = 4\sqrt{2} = r$$

$$\theta = \tan^{-1}(1/1) = 45° \text{ or } \pi/4 \text{ radians}$$

$$4+4i = 4\sqrt{2}\,(\cos\theta+i\sin\theta)$$

Since $\sin 45° = 1/\sqrt{2}$ it is easily seen that the polar form is equal to $4+4i$.

6.3 Complex conjugates

The complex conjugate of z, usually written as z^*, is obtained simply by reversing the sign of the imaginary part of z. Thus $(x-iy)$ is the complex conjugate of $(x+iy)$. When any complex number is multiplied by its complex conjugate the product is a real number,

e.g. $$(x+iy)(x-iy) = x^2+y^2 = zz^*$$

Note: zz^* is always a real number.

Example. Simplify

$$\frac{2+i}{3-i}$$

$$\frac{2+i}{3-i} = \frac{(2+i)(3+i)}{(3-i)(3+i)} = \frac{6+5i+i^2}{9-i^2}$$

$$= \frac{5+5i}{10} = \tfrac{1}{2}(1+i)$$

Note that the normal algebraic rules are used.

Complex conjugates are widely used in quantum mechanics. Wave functions are often complex, and to evaluate probabilities integrals of the product of the wave function and its complex conjugate are evaluated—the probability will therefore be a real number.

6.4 Euler's formula

Euler's formula states,

$$e^{i\theta} = \cos\theta + i\sin\theta$$

Using this formula, a complex number may be expressed as

$$z = re^{i\theta}$$

This allows us to write trigonometric functions such as $\sin\theta$ or $\cos\theta$ in terms of complex numbers. This is useful since it simplifies the evaluation of integrals containing sin and cos.

$$e^{i\theta} = \cos\theta + i\sin\theta$$

$$e^{-i\theta} = \cos\theta - i\sin\theta$$

since $$\cos(-\theta) = \cos\theta$$

and $$\sin(-\theta) = -\sin\theta$$

Thus $$e^{i\theta}+e^{-i\theta} = 2\cos\theta$$

therefore $$\cos\theta = \frac{e^{i\theta}+e^{-i\theta}}{2}$$

and by subtraction

$$\sin\theta = \frac{e^{i\theta}-e^{-i\theta}}{2i}$$

Example. Write the complex number $4+4i$ in exponential form.

From a previous example

$$r = 4\sqrt{2}$$

$$\theta = \pi/4 \text{ radians}$$

$$4+4i = 4\sqrt{2}e^{i\pi/4}$$

since $$z = re^{i\theta}$$

Note that θ is expressed in radians.

Example. What is the value of $3e^{-i\pi/2}$?

$$\begin{aligned} 3e^{-i\pi/2} &= 3(\cos(-\pi/2)+i\sin(-\pi/2)) \\ &= 3(\cos\pi/2-i\sin\pi/2) \\ &= 3(0-i(1)) \\ &= -3i \end{aligned}$$

Example. What is the value of $(1+i)^8$?

$$z = 1+i; \quad r = \sqrt{2}; \quad \theta = \pi/4$$

$$z = \sqrt{2}e^{i\pi/4}$$

$$\begin{aligned} z^8 &= 16e^{2\pi i} \\ &= 16(\cos 2\pi+i\sin 2\pi) \\ &= 16(1+i(0)) \\ &= 16 \end{aligned}$$

This could have been evaluated by multiplying out $(1+i)(1+i)(1+i)\ldots$ but the use of the exponential form greatly simplifies the problem.

Example. Prove that $\sin^2 z+\cos^2 z = 1$

$$\sin^2 z = \left\{\frac{e^{iz}-e^{-iz}}{2i}\right\}^2 = \frac{e^{2iz}-2+e^{-2iz}}{-4}$$

$$\cos^2 z = \left\{\frac{e^{iz}+e^{-iz}}{2}\right\}^2 = \frac{e^{2iz}+2+e^{-2iz}}{4}$$

therefore $$\sin^2 z+\cos^2 z = \tfrac{2}{4}+\tfrac{2}{4} = 1$$

6.5 Hyperbolic functions

If we derive $\sin z$ and $\cos z$ for a pure imaginary number z, i.e. $z = iy$, then we have

$$\sin iy = \frac{e^{-y}-e^{y}}{-2}$$

and $$\cos iy = \frac{e^{-y}+e^{y}}{2}$$

The real functions on the right are known as *hyperbolic functions* and the hyperbolic sine, written *sinh*, and the hyperbolic cosine written *cosh*, are defined

$$\sinh z = \frac{e^{z}-e^{-z}}{2}$$

$$\cosh z = \frac{e^{z}+e^{-z}}{2}$$

and the hyperbolic tangent, *tanh*, is given by

$$\tanh z = \frac{e^{z}-e^{-z}}{e^{z}+e^{-z}}$$

by analogy with the relationship

$$\tan z = \sin z/\cos z$$

(see Osborne's Rule, Appendix A6)

7 *Vector Analysis*

7.1 Introduction

Many quantities have both direction and magnitude e.g. the electrostatic attraction of an electron towards a nucleus, force at a point, velocity, acceleration, etc., and these may be conveniently expressed as *vectors.* Other quantities, which have only magnitude, e.g. time, mass, temperature, are known as *scalars.*

A vector may be represented by an arrow, the length of the arrow giving the magnitude, and the direction of the arrow representing direction of the vector.

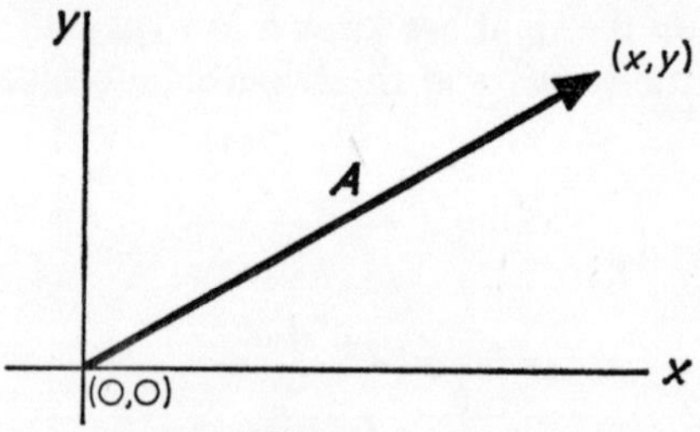

Fig. 7.1

The vector $\boldsymbol{A}$ is drawn in two dimensions.† The magnitude of $\boldsymbol{A}$, $|\boldsymbol{A}|$, is equal to the length of the vector,

$$|\boldsymbol{A}| = \sqrt{(x^2+y^2)}$$

or in three dimensions

$$|\boldsymbol{A}| = \sqrt{(x^2+y^2+z^2)}$$

If we have a set of orthogonal axes and $\boldsymbol{p}$ is a vector of unit length in the positive x-direction, and $\boldsymbol{q}$ and $\boldsymbol{r}$ unit vectors in the positive y- and z-directions, then any vector may be written in terms of its components and the unit vectors (Fig. 7.2).

e.g. $$\boldsymbol{A} = x\boldsymbol{p}+y\boldsymbol{q}+z\boldsymbol{r}$$

---

† In script a vector is often denoted by an arrow or half-arrow placed above the symbol e.g. $\vec{A}$, $\vec{a}$ or $\overrightarrow{OP}$.

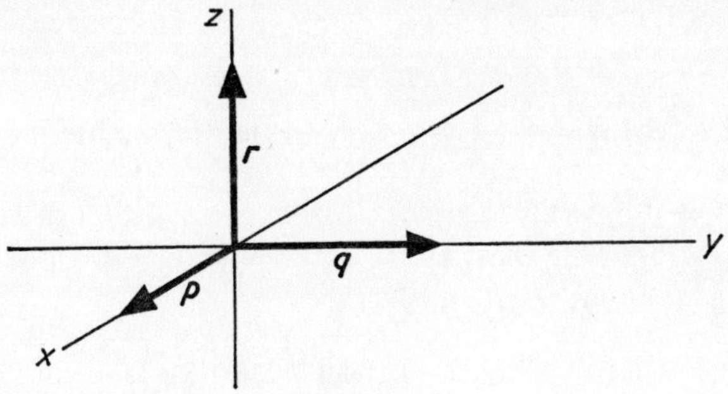

Fig. 7.2

It is common practice to refer to a vector simply by its components (x, y, z) e.g. the vector $(3, -2, 1)$.

7.2 Addition of vectors

Vectors may be added or subtracted using the normal algebraic laws,

i.e. $$\boldsymbol{A}+\boldsymbol{B} = \boldsymbol{B}+\boldsymbol{A}$$

$$(\boldsymbol{A}+\boldsymbol{B})+\boldsymbol{C} = \boldsymbol{A}+(\boldsymbol{B}+\boldsymbol{C})$$

This may be done graphically where the vectors are added using the parallelogram law, as in Fig. 7.3.

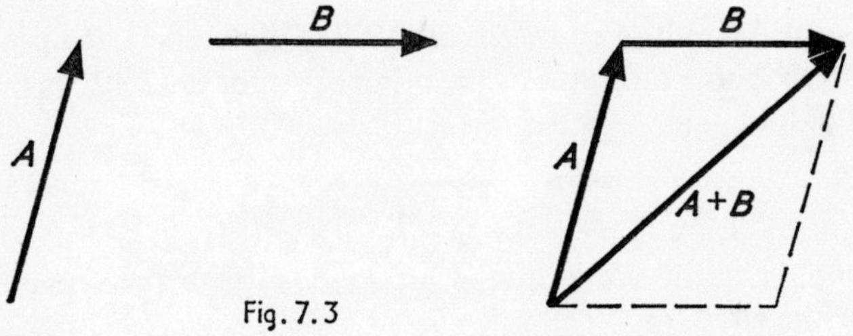

Fig. 7.3

The resultant vector $\boldsymbol{A}+\boldsymbol{B}$ is the diagonal of a parallelogram based on $\boldsymbol{A}$ and $\boldsymbol{B}$.

$-\boldsymbol{B}$ has the same magnitude as $\boldsymbol{B}$ but the opposite direction so that $\boldsymbol{A}-\boldsymbol{B}$ may be represented as shown in Fig. 7.4.

If however we know the components of the vectors, $(x\boldsymbol{p}+y\boldsymbol{q}+z\boldsymbol{r})$ then addition or subtraction may be done by adding or subtracting the components.

If $\quad A = x_1 p + y_1 q + z_1 r$

and $\quad B = x_2 p + y_2 q + z_2 r$

$$A+B = (x_1+x_2)p+(y_1+y_2)q+(z_1+z_2)r$$

Example. Find the resultant vector D from the addition of vectors A, B and C with components (1, 1, 1), (2, 1, 3) and (−1, −2, 1).

$$\begin{aligned} D &= A+B+C \\ &= (1+2-1,\ 1+1-2,\ 1+3+1) \\ &= (2, 0, 5) \end{aligned}$$

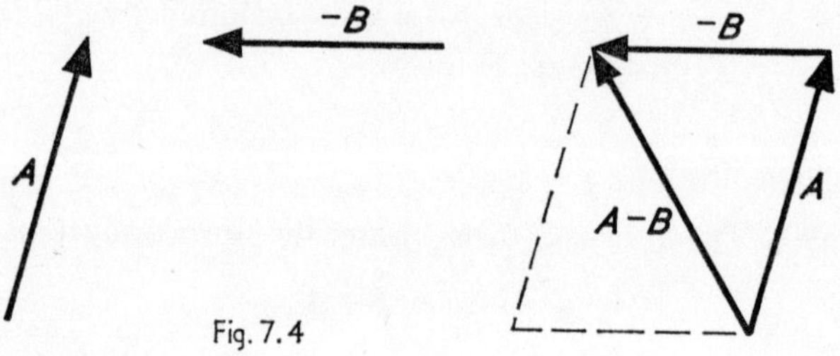

Fig. 7.4

7.3 Multiplication of vectors

(i) *Multiplication of vector and scalar*

When a vector is multiplied by a scalar, N (i.e. a number), the product is a vector with the same direction as the original vector but with its magnitude increased N-fold, e.g.

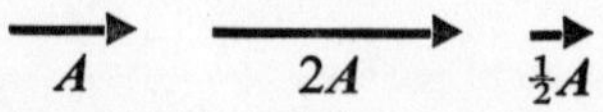

If the components of the vector A are (x, y, z) then the components of vector NA are (Na, Ny, Nz).

(ii) *Multiplication of two vectors*

Vectors may be multiplied together in two distinct ways:

(a) to give a scalar product (i.e. a number) or

(b) to give a vector product (i.e. a vector).

It is perhaps useful to visualise the results of these different types of multiplication before dealing with them in detail. The scalar product is simply a number and may be visualised in terms of projections of one vector upon another. For example, if we have two vectors, $\boldsymbol{A}$ and $\boldsymbol{B}$, the scalar product $\boldsymbol{A}.\boldsymbol{B}$ may be written as

$$\boldsymbol{A}.\boldsymbol{B} = |A|\,|B| \cos \theta.$$

$\boldsymbol{B} \cos \theta$ = projection of $\boldsymbol{B}$ upon $\boldsymbol{A}$ i.e. x in Fig. 7.5(a). So the scalar product $\boldsymbol{A}.\boldsymbol{B}$ is equal to $|A|$ multiplied by the projection of $\boldsymbol{B}$ upon $\boldsymbol{A}$ and it is seen that $\boldsymbol{A}.\boldsymbol{B} = \boldsymbol{B}.\boldsymbol{A}$.

The vector product $\boldsymbol{A} \times \boldsymbol{B}$, on the other hand, is itself another vector, not a number, and is perpendicular to the two original vectors (Fig. 7.5(b). The direction of the vector product has to be defined. $\boldsymbol{A} \times \boldsymbol{B}$ is a vector in the opposite direction to $\boldsymbol{B} \times \boldsymbol{A}$.

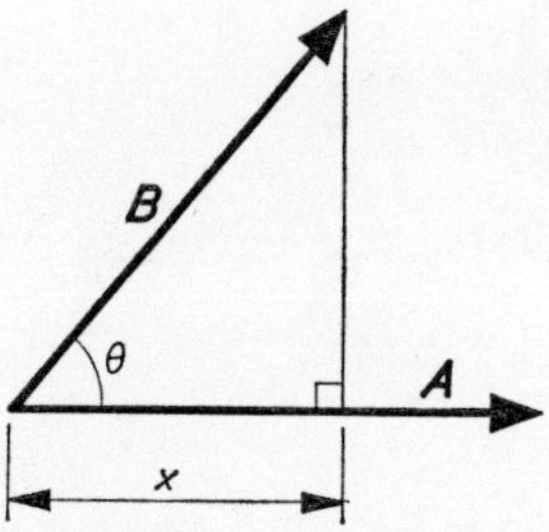

Fig. 7.5(a)

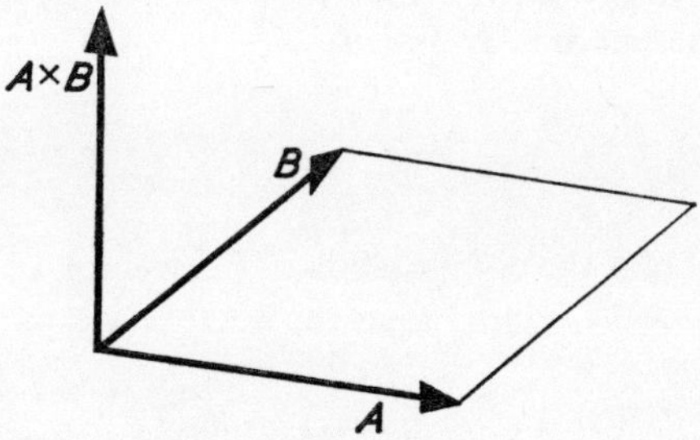

Fig. 7.5(b)

We shall now consider these two types of vector multiplication in more detail.

(a) *Scalar (dot) product*

The scalar product is defined as the product of the magnitude of $\boldsymbol{A}$, $|A|$, the magnitude of $\boldsymbol{B}$, $|B|$, and the cosine of the angle between $\boldsymbol{A}$ and $\boldsymbol{B}$. This type of product is written as $\boldsymbol{A}.\boldsymbol{B}$.

$$\boldsymbol{A}.\boldsymbol{B} = |A|\,|B| \cos \theta$$

If, however, we know the components of the vectors then we can determine the scalar product from the relationship

$$\boldsymbol{A}.\boldsymbol{B} = x_1x_2+y_1y_2+z_1z_2$$

Example. Find the angle between the vectors $\boldsymbol{A}$, (3, 6, 9) and $\boldsymbol{B}$, (−2, 3, 1)

$$|A| = \sqrt{(3^2+6^2+9^2)} = 3\sqrt{14}$$

$$|B| = \sqrt{14}$$

$$\text{Now} \qquad \boldsymbol{A}.\boldsymbol{B} = 3(-2)+6(3)+9(1)$$
$$= 21$$
$$\text{therefore} \qquad 21 = (3\sqrt{14})(\sqrt{14})(\cos\theta)$$
$$\cos\theta = 0{\cdot}5$$
$$\theta = 60^\circ$$

Two special cases arise: (i) where the vectors are mutually perpendicular, when $\theta = 90^\circ$, $\cos\theta = 0$ and the scalar product, $\boldsymbol{A}.\boldsymbol{B}$ is zero; and (ii) where the vectors are parallel, when $\theta = 0^\circ$, $\cos\theta = 1$ and $\boldsymbol{A}.\boldsymbol{B} = |\boldsymbol{A}|\,|\boldsymbol{B}|$.

Thus if $x_1x_2+y_1y_2+z_1z_2 = 0$ then the two vectors must be mutually perpendicular and if

$$\frac{x_1}{x_2} = \frac{y_1}{y_2} = \frac{z_1}{z_2}$$

(i.e. if the components of one vector are a multiple of the components of a second vector) then the vectors are parallel.

(b) *Vector (cross) product*

The second type of multiplication to give the vector (or cross) product is written as $\boldsymbol{A}\times\boldsymbol{B}$.

This is defined as giving a vector with magnitude

$$|\boldsymbol{A}\times\boldsymbol{B}| = |\boldsymbol{A}|\,|\boldsymbol{B}|\sin\theta,$$

θ being the angle ($\leq 180^\circ$) between vectors $\boldsymbol{A}$ and $\boldsymbol{B}$. We still require to know the direction of this vector $\boldsymbol{A}\times\boldsymbol{B}$ and this is defined as being perpendicular to the plane containing vectors $\boldsymbol{A}$ and $\boldsymbol{B}$. We must further indicate the direction of the vector product. This can be understood by considering Fig. 7.6.

In Fig. 7.6(a) the vector products $\boldsymbol{A}\times\boldsymbol{B}$ and $\boldsymbol{B}\times\boldsymbol{A}$ are drawn in the correct directions. Note that their magnitudes are identical but that they are opposite in direction.

To determine the defined direction of a vector product the following rule may be used. If the right hand index finger points in the direction of $\boldsymbol{A}$ and the second finger indicates the shortest way to $\boldsymbol{B}$ then the thumb will define the direction of $\boldsymbol{A}\times\boldsymbol{B}$, perpendicular to the plane containing $\boldsymbol{A}$ and $\boldsymbol{B}$ (see Fig. 7.6(b)).

With vector products the normal algebraic laws of multiplication (e.g. $xy = yx$) *do not hold* since

$$\boldsymbol{A}\times\boldsymbol{B} \neq \boldsymbol{B}\times\boldsymbol{A}$$

i.e. vector multiplication is not commutative.

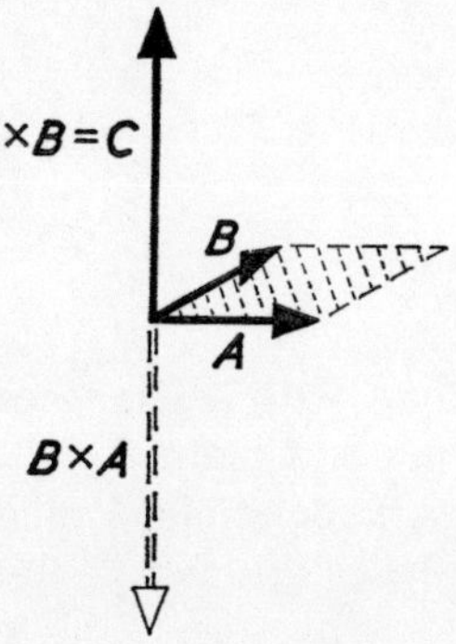

Fig. 7.6 (a)

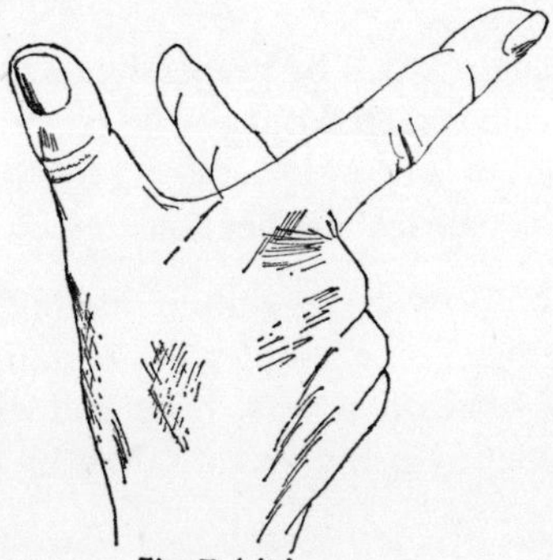

Fig. 7.6 (b)

If the components of vectors $\boldsymbol{A}$ and $\boldsymbol{B}$ are known,

$$\boldsymbol{A} = x_1\boldsymbol{p}+y_1\boldsymbol{q}+z_1\boldsymbol{r}, \quad \boldsymbol{B} = x_2\boldsymbol{p}+y_2\boldsymbol{q}+z_2\boldsymbol{r}$$

then $\boldsymbol{A}\times\boldsymbol{B}$ may be found from the determinant (see Section 8.2).

$$\boldsymbol{A}\times\boldsymbol{B} = \begin{vmatrix} \boldsymbol{p} & \boldsymbol{q} & \boldsymbol{r} \\ x_1 & y_1 & z_1 \\ x_2 & y_2 & z_2 \end{vmatrix}$$

i.e. $\boldsymbol{A}\times\boldsymbol{B} = (y_1z_2-y_2z_1)\boldsymbol{p}-(x_1z_2-x_2z_1)\boldsymbol{q}+(x_1y_2-x_2y_1)\boldsymbol{r}$

Example. Classically, the angular momentum about a point O is a vector, $\boldsymbol{L}$, which is defined as the vector product of $\boldsymbol{s}$, the distance from the point, and the linear momentum $\boldsymbol{P}$ (Fig. 7.7).

$$\boldsymbol{L} = \boldsymbol{s}\times\boldsymbol{P}$$

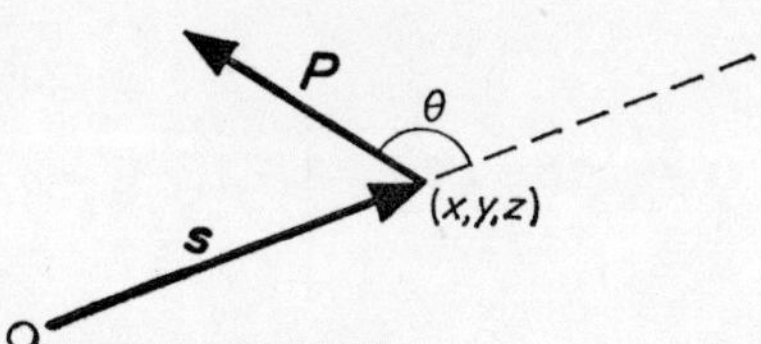

Fig. 7.7

The vector $\boldsymbol{L}$ can be written

$$\boldsymbol{L} = \begin{vmatrix} \boldsymbol{p} & \boldsymbol{q} & \boldsymbol{r} \\ x & y & z \\ p_x & p_y & p_z \end{vmatrix}$$

where (x, y, z) and (p_x, p_y, p_z) are the components of the vectors $\boldsymbol{s}$ and $\boldsymbol{P}$ respectively.

The vector $\boldsymbol{L}$ will be perpendicular to the paper (Fig. 7.7) and by the right hand rule directed out of the paper.

7.4 Geometrical applications

Vectors provide a simple method of dealing with some geometrical problems. For example, the equations of lines and planes, and the distances between points, lines and planes may be determined much more conveniently using vectors than by using analytical geometry.

We have already seen how the scalar product allows us to find the angle between two vectors. The *vector* product on the other hand was defined as a vector normal to the plane of the two original vectors. If the vector product is $a\boldsymbol{p}+b\boldsymbol{q}+c\boldsymbol{r}$ then knowing one point in the plane, say (x_0, y_0, z_0) we can write the equation of the original plane using the identity

$$a(x-x_0)+b(y-y_0)+c(z-z_0) = 0$$

Example. Find the equation of the plane through the three points $A(1, 2, 2)$, $B(-1, 2, 1)$ and $C(2, 1, 1)$.

A vector joining any two of these points will be in the plane and can be written, in general, as

$$\boldsymbol{V} = \boldsymbol{p}(x_2-x_1)+\boldsymbol{q}(y_2-y_1)+\boldsymbol{r}(z_2-z_1)$$

where $\boldsymbol{V}$ is the vector joining (x_1, y_1, z_1) to (x_2, y_2, z_2)

Thus

$$\boldsymbol{AB} = -2\boldsymbol{p}-\boldsymbol{r}$$

$$\boldsymbol{BC} = 3\boldsymbol{p}-\boldsymbol{q}$$

The cross product of these two vectors, $\boldsymbol{N}$, will be normal to the plane and may be written

$$\boldsymbol{N} = (\boldsymbol{AB}\times\boldsymbol{BC}) = \begin{vmatrix} \boldsymbol{p} & \boldsymbol{q} & \boldsymbol{r} \\ -2 & 0 & -1 \\ 3 & -1 & 0 \end{vmatrix}$$

$$= -1\boldsymbol{p}-3\boldsymbol{q}+2\boldsymbol{r}$$

Now using the point (1, 2, 2), say, the plane containing all three points is then seen to be

$$-1(x-1)-3(y-2)+2(z-2) = 0$$

i.e. $$x+3y-2z = 3$$

7.5 n-Dimensional vectors

We have so far considered vectors only in two or three dimensional space, i.e. those vectors which it is possible to visualise.

The vector concept can however be extended to a space with more than three dimensions. For example, it requires six variables to specify the dynamics of a particle: x, y and z to give its position and P_x, P_y, P_z to denote the components of its momentum. These six variables (x, y, z, P_x, P_y, P_z) may be interpreted as a point in six dimensional space or as the components of a vector in six dimensional space. It is impossible for us to visualise space with more than three dimensions but this does not preclude the use of such space in mathematical terms.

7.6 Differentiation of a vector function with respect to a scalar variable

Let $A(u)$ be a vector function of a scalar variable u. Just as a continuous scalar function of u, $f(u)$, can be differentiated, so if $A(u)$ is continuous at u, $A(u)$ can be differentiated with respect to u.

$$\frac{dA}{du} = \lim_{h\to 0} \frac{A(u+h)-A(u)}{h}$$

where h is a small increment in u.

The components of A along the orthogonal axes will also be functions of u

i.e. $$A(u) = a_x(u)\boldsymbol{p}+a_y(u)\boldsymbol{q}+a_z(u)\boldsymbol{r}$$

where $\boldsymbol{p}$, $\boldsymbol{q}$, $\boldsymbol{r}$ are the unit vectors in the positive x, y, z directions respectively.

Thus $$\frac{dA}{du} = \frac{da_x}{du}\boldsymbol{p}+\frac{da_y}{du}\boldsymbol{q}+\frac{da_z}{du}\boldsymbol{r}.$$

Example. If a particle moves along a curve so that its position vector is

$$\boldsymbol{s} = \cos nt\boldsymbol{p}+\sin nt\boldsymbol{q}+t\boldsymbol{r}$$

where t is the time variable, calculate the velocity and acceleration vectors of the particle.

If δs is the change in s during the time interval δt, $\delta s/\delta t$ represents the average rate of change of s during this time interval. As δt tends to zero, this will represent the velocity of the particle relative to the origin at time t.

Thus the velocity vector, v, will be given by

$$\lim_{0} \frac{\delta s}{\delta t} = \frac{ds}{dt} = -n \sin nt\boldsymbol{p} + n \cos nt\boldsymbol{q} + \boldsymbol{r}$$

Similarly the acceleration vector

$$\boldsymbol{a} = \frac{d\boldsymbol{v}}{dt} = \frac{d^2\boldsymbol{s}}{dt^2}$$

$$= -n^2 \cos nt\boldsymbol{p} - n^2 \sin nt\boldsymbol{q}$$

If the vector $A(u, v, w, \ldots z)$ is a function of several independent variables, $u, v, w, \ldots z$, then the partial derivative of A with respect to one of these, say u, is written

$$\frac{\partial A}{\partial u} \quad \text{or} \quad \left(\frac{\partial A}{\partial u}\right)_{v, w \ldots z}$$

and represents the rate of change of A with respect to u, the other independent variables remaining constant.

If A and B are vector functions of u, the following rules are easily proved.

(i) $\dfrac{d}{du} C = 0$ where C is independent of u.

(ii) $\dfrac{d}{du}(A+B) = \dfrac{dA}{du} + \dfrac{dB}{du}$

(iii) $\dfrac{d}{du}(fA) = f\dfrac{dA}{du} + \dfrac{df}{du}A$ where f is a function of u.

(iv) $\dfrac{d}{du}(A.B) = A.\dfrac{dB}{du} + \dfrac{dA}{du}.B.$

(v) $\dfrac{d}{du}(A \times B) = A \times \dfrac{dB}{du} + \dfrac{dA}{du} \times B.$

Note that in (v) the order of multiplication must be retained.

7.7 Integration of vector functions

If a continuous vector function of u, $\boldsymbol{A}(u)$, has a differential coefficient $\boldsymbol{B}(u)$

$$\frac{\mathrm{d}\boldsymbol{A}}{\mathrm{d}u} = \boldsymbol{B}(u)$$

then $\boldsymbol{A}$ is the indefinite integral of $\boldsymbol{B}$ with respect to u

i.e. $\int \boldsymbol{B}(u)\,\mathrm{d}u = \boldsymbol{A}(u)+\boldsymbol{C}$ where $\boldsymbol{C}$ is a constant vector.

7.8 Scalar and vector fields

Physical quantities often have different values at different points within a given space, e.g. the variation of temperature in a hall, the velocity of flow of water in a river, the gravitational field of the earth or the electric field of a point charge. At any point within the given space, the physical property has a definite value. The term 'field' is taken to mean both the region involved and the value of the property within the region. Sometimes the term *point function* is used instead of field. The physical quantity may be scalar, e.g. temperature, in which case we have a *scalar field* or *scalar point function* or a vector, e.g. velocity of flow, in which case we have a *vector field* or *vector point function.*

Within such a field, a curve (in a two dimensional field) or a surface (in a three dimensional field) on which all points have the same value of the physical quantity in question is called an *equipotential.* For example, the equipotential surfaces in the electrostatic field of a point charge are spheres with centres at the point charge. Equipotentials are analogous to the lines on a contour map. Some have specific names e.g. 'isothermals' for temperature equipotentials.

A scalar field is represented by a function $\phi(x, y, z)$ giving the value of the scalar ϕ at any point (x, y, z) in the region. If the value of the scalar is time dependent, the function $\phi\ (x, y, z, t)$ would represent the field. For example a time independent scalar field might be represented by $3xy+z^2$. Thus at each point (x_1, y_1, z_1), (x_2, y_2, z_2) etc., the value of the scalar can be determined.

A vector field is represented by $\boldsymbol{F}\ (x, y, z)$ in a given region and, at any point (x, y, z) in the region, the vector is expressed in component form

$$\boldsymbol{F} = F_x\boldsymbol{p}+F_y\boldsymbol{q}+F_z\boldsymbol{r}$$

where F_x, F_y, F_z are functions of x, y, z. For example, $x^2\boldsymbol{p}+3xy\boldsymbol{q}$ represents a two dimensional vector field and $xyz\boldsymbol{p}+x^2\boldsymbol{q}+yz\boldsymbol{r}$ represents a three dimensional vector field.

7.9 Line integrals

Let us consider a curve joining points M and N (Fig. 7.8(a)). If the components of each point on the curve are functions of u, $x = x(u)$, $y = y(u)$ and $z = z(u)$, and u increases monotonically from M ($u = u_m$) to N($u = u_n$) then the integral

$$\int_{u_m}^{u_n} \phi(x(u), y(u), z(u))\, du$$

(where $\phi\ (x, y, z)$ is a scalar function of position for all points on the curve between M and N) is called the *line integral* of ϕ.

If the curve is not a continuous function of u between M and N then the line integral is the sum of the line integrals of each segment (Fig. 7.8(b))

i.e. $$\int_{u_m}^{u_l} \phi'\, du + \int_{u_l}^{u_n} \phi''\, du.$$

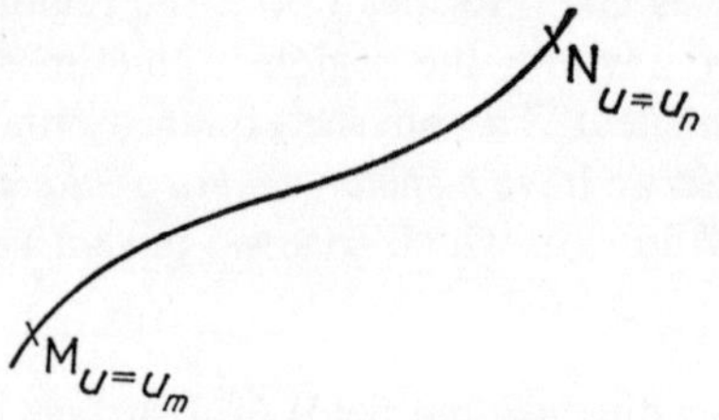

Fig. 7. 8(a)

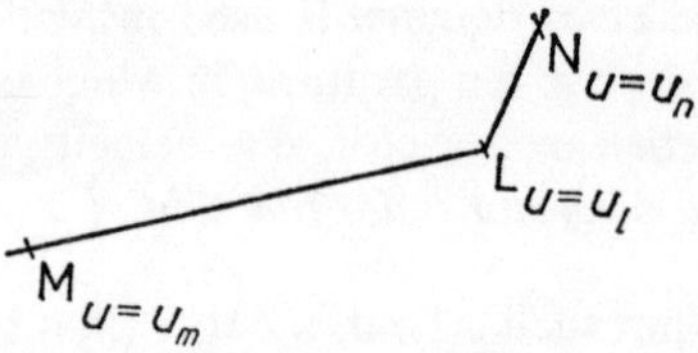

Fig. 7. 8 (b)

If the curve in Fig. 7.8(a) is described in the opposite direction, i.e. N to M then the line integral with respect to u is defined as

$$\int_{u_n}^{u_m} \phi\, du = -\int_{u_m}^{u_n} \phi\, du.$$

If M and N coincide and integration is to be carried out over the closed loop C (Fig. 7.9) then the notation

$$\oint_C \phi\, du$$

is used for the line integral (C denotes integration over a curve and $\oint$ denotes a closed loop). It is assumed that the loop is described in the direction of increasing u.

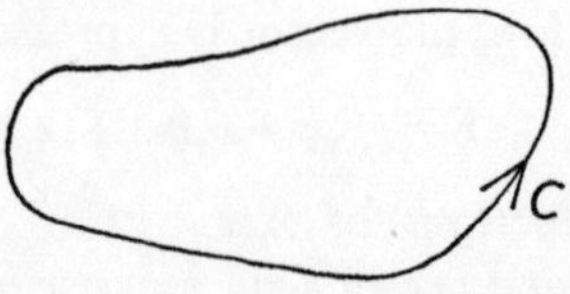

Fig. 7.9

The line integral of a vector function $\boldsymbol{F}(x, y, z)$, where the components F_x, F_y, F_z, are all functions of u, from $M(u = u_m)$ to $N(u = u_n)$ is

$$\boldsymbol{p}\int_{u_m}^{u_n} F_x(u)\,\mathrm{d}u + \boldsymbol{q}\int_{u_m}^{u_n} F_y(u)\,\mathrm{d}u + \boldsymbol{r}\int_{u_m}^{u_n} F_z(u)\,\mathrm{d}u$$

assuming that u increases monotonically from u_m to u_n.

One of the most important types of line integral is the *tangential line integral* (T.L.I.) of a vector. This enables us to calculate, for example, the work done by a force $\boldsymbol{F}$ operating on a particle in moving it from M to N along the curve in Fig. 7.10.

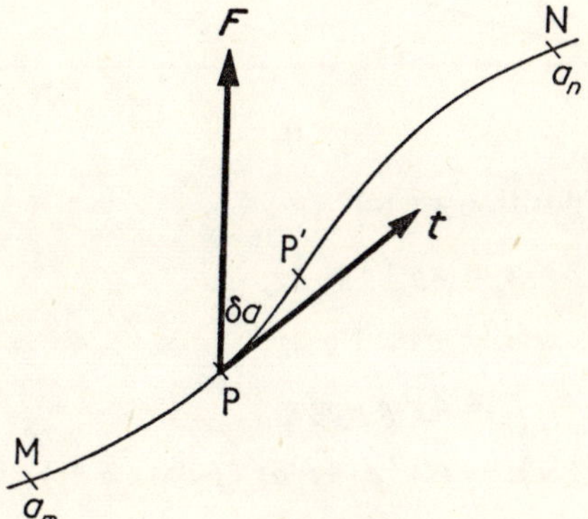

Fig. 7.10

If the length of the arc of the curve PP' is δa then the work done in moving the particle from P to P' is

$$\boldsymbol{F}.\boldsymbol{t}\delta a$$

(where $\boldsymbol{t}$ is the unit tangential vector at P in the direction of increasing a) because $\boldsymbol{F}.\boldsymbol{t}$ is the component of $\boldsymbol{F}$ in the direction of $\boldsymbol{t}$. Thus if the curve from M to N is divided into segments δa the total work done in moving the particle from M to N is approximately

$$\sum_{\text{all } \delta a} \boldsymbol{F}.\boldsymbol{t}\delta a$$

As $\delta a \to 0$ $$\lim_{\delta a \to 0} \sum_{\delta a} \boldsymbol{F}.\boldsymbol{t}\delta a = \int_{a_m}^{a_n} \boldsymbol{F}.\boldsymbol{t}\,\mathrm{d}a$$

If $\boldsymbol{\alpha}\,(x, y, z)$ is the position vector of P, we can replace $\boldsymbol{t}\delta a$ by $\delta\boldsymbol{\alpha}$ (i.e. $\boldsymbol{PP'}$ in Fig. 7.10) and in the limit as $\delta\boldsymbol{\alpha} \to 0$

the work done $$= \int_{M\,C}^{N} \boldsymbol{F}.\mathrm{d}\boldsymbol{\alpha}$$

where $\mathrm{d}\boldsymbol{\alpha}$ is a differential vector whose components are $\mathrm{d}x$, $\mathrm{d}y$, $\mathrm{d}z$. Note that $\boldsymbol{F}.\mathrm{d}\boldsymbol{\alpha}$ is a scalar and integration is therefore straightforward.

Example. If a force $\boldsymbol{F} = 4y^2\boldsymbol{p}-xy\boldsymbol{q}$ acts on a particle on (i) a straight line path, (ii) a parabolic path and (iii) a path $(0, 0) \rightarrow (1, 1) \rightarrow (2, 1)$ from points (0, 0) to (2, 1) (see Fig. 7.11) calculate the work done in each case.

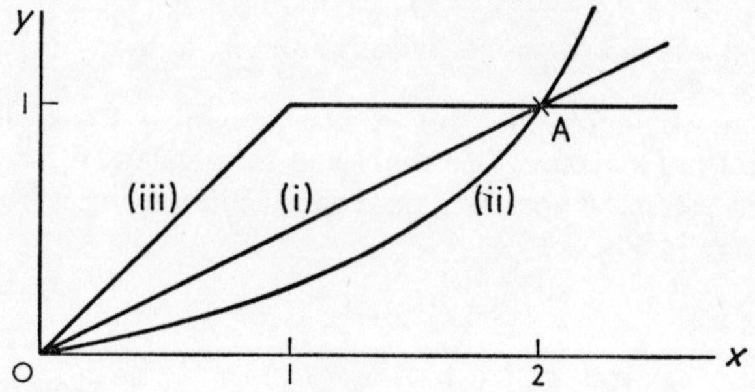

Fig. 7.11

In a plane the position vector

$$\boldsymbol{\alpha} = x\boldsymbol{p}+y\boldsymbol{q}$$

therefore $$\mathrm{d}\boldsymbol{\alpha} = \boldsymbol{p}\,\mathrm{d}x+\boldsymbol{q}\,\mathrm{d}y$$

Since $$\boldsymbol{F} = 4y^2\boldsymbol{p}-xy\boldsymbol{q}$$

$$\boldsymbol{F}.\mathrm{d}\boldsymbol{\alpha} = (4y^2\boldsymbol{p}-xy\boldsymbol{q}).(\boldsymbol{p}\,\mathrm{d}x+\boldsymbol{q}\,\mathrm{d}y)$$

$$= 4y^2\,\mathrm{d}x-xy\,\mathrm{d}y$$

and the work done $$= \int_0^A (4y^2\,\mathrm{d}x-xy\,\mathrm{d}y)$$

We have now to write the integrand in terms of one variable.

In case (i), the equation of the line OA is

$$y = x/2$$

therefore $$\mathrm{d}y = \tfrac{1}{2}\,\mathrm{d}x$$

Substituting for y and $\mathrm{d}y$ we obtain the following expression

work done $$= \int_0^2 \left(\frac{4x^2}{4}\,\mathrm{d}x-\frac{x^2}{2}\tfrac{1}{2}\,\mathrm{d}x\right)$$

$$= \int_0^2 \frac{3x^2}{4}\,\mathrm{d}x$$

$$= \left[\frac{x^3}{4}\right]_0^2$$

$$= 2$$

In case (ii) the equation of the curve is $y = x^2/4$

therefore $\qquad dy = \frac{1}{2}x\,dx$

Substituting we obtain

$$\text{work done} = \int_0^2 \left(4\frac{x^4}{16} - x\frac{x^2}{4}\frac{x}{2}\right) dx$$

$$= \int_0^2 \frac{x^4}{8}\,dx$$

$$= \left[\frac{x^5}{40}\right]_0^2$$

$$= 32/40$$

In case (iii) the curve is not continuous and the work done is the sum of the work done in moving the particle from (0, 0) to (1, 1) and that done in moving the particle from (1, 1) to (2, 1).

The equation of the straight line joining (0, 0) and (1, 1) is

$$y = x$$

and $\qquad dy = dx$

$$\text{i.e. work done} = \int_0^1 (4y^2\,dy - y^2\,dy)$$

$$= \int_0^1 3y^2\,dy$$

$$= [y^3]_0^1$$

$$= 1$$

The equation of the straight line joining (1, 1) and (2, 1) is

$$y = 1$$

and $\qquad dy = 0$

$$\text{therefore work done} = \int_1^2 4\,dx - 0$$

$$= [4x]_1^2$$

$$= 4$$

therefore work done in moving the particle from O to P

along path (iii) $= 4+1 = 5$.

It is seen that, in this example, the work done in moving the particle from the origin to A along the different paths is not the same. This is not always the case. If $\int_{M}^{N}{}_{C} \boldsymbol{F}.\mathrm{d}\boldsymbol{\alpha}$ is the same between M and N regardless of the path chosen, the field in question is called a *conservative field.*

7.10 Surface and volume integrals

Integration over a surface or a volume is similar to that described above for a curve.

A surface integral is a double integral and it is written

$$\int_S \phi \, \mathrm{d}S \quad \text{or} \quad \iint_S \phi \, \mathrm{d}S \equiv \iint_S \phi(u, v) \, du \, dv.$$

A volume integral is a triple integral and it is written

$$\int_V \phi \, \mathrm{d}V \quad \text{or} \quad \iiint_V \phi \, \mathrm{d}V \equiv \iiint_V \phi(u, v, w) \, du \, dv \, dw.$$

7.11 The gradient vector (grad) of a scalar field

The vector differential operator del (∇) is defined as

$$\nabla = \boldsymbol{p}(\partial/\partial x)+\boldsymbol{q}(\partial/\partial y)+\boldsymbol{r}(\partial/\partial z).$$

It may operate on either scalar or vector fields. In this section we will consider the significance of the operation on a scalar field. If $\phi\,(x, y, z)$ is a scalar field which has first partial derivatives with respect to x, y, z then

$$\nabla.\phi = \boldsymbol{p}(\partial\phi/\partial x)+\boldsymbol{q}(\partial\phi/\partial y)+\boldsymbol{r}(\partial\phi/\partial z)$$

is called the gradient of ϕ (grad ϕ).

Grad ϕ is a vector whose direction is that of the greatest rate of increase of ϕ with respect to displacement and whose magnitude is equal to that rate of increase.

An analogy is that of a climber standing part-way up a mountain and observing the direction of the steepest slope towards the summit. It is obviously useful in a scalar field to be able to determine the gradient vector at a particular point. Thus to find the direction with the greatest rate of increase in ϕ, grad ϕ is evaluated, and conversely for the greatest rate of decrease in ϕ, $-$grad ϕ is evaluated.

Example. Consider the flow of solute through a solution and suppose that only solute molecules diffuse. We will represent this flow by $\boldsymbol{J}_s$. Assume also that the velocity of diffusion of the solute can be represented by $\boldsymbol{v}_s$ which is proportional to the driving force for diffusion i.e. the negative gradient of the chemical potential of the solute, μ_s.

$$\text{i.e. } \boldsymbol{v}_s = -\omega_s \operatorname{grad} \mu_s.$$

where ω_s is the mobility of the solute in the medium. In ideal solutions the chemical potential is related to the molar concentration, c_s by the expression

$$\mu_s = \mu_s^0 + RT \ln c_s$$

where μ_s^0 is the standard chemical potential of the solute.

$$\text{Grad } \mu_s = \nabla\mu_s = \nabla\mu_s^0 + \frac{RT}{c_s}[\boldsymbol{p}(\partial c_s/\partial x) + \boldsymbol{q}(\partial c_s/\partial y) + \boldsymbol{r}(\partial c_s/\partial z)]$$

since c_s is a function of x, y, z and

$$(\partial/\partial x) \ln c_s = \frac{1}{c_s}(\partial/\partial x)(c_s) \qquad \text{(see Section 2.4)}$$

i.e.
$$\operatorname{grad} \mu_s = 0 + \frac{RT}{c_s} \operatorname{grad} c_s$$

i.e.
$$\boldsymbol{v}_s = -\frac{RT\omega_s}{c_s} \operatorname{grad} c_s$$

and
$$\boldsymbol{J}_s = c_s \boldsymbol{v}_s$$

therefore
$$\boldsymbol{J}_s = -RT\omega_s \operatorname{grad} c_s.$$

This is Fick's first law of diffusion: $\boldsymbol{J}_s = -D \operatorname{grad} c_s$ where the diffusion coefficient $D = RT\omega_s$.

The following identities may be useful. If $u(x, y, z)$ and $v(x, y, z)$ are differentiable scalar fields and $\phi(u)$ is a differentiable function of u,

(i) $\operatorname{grad}(u+v) = \operatorname{grad} u + \operatorname{grad} v$

(ii) $\operatorname{grad}(uv) = u \operatorname{grad} v + v \operatorname{grad} u$

(iii) $\operatorname{grad} \phi(u) = (\mathrm{d}\phi/\mathrm{d}u) \operatorname{grad} u.$

It is also useful to note that for a conservative vector field $\boldsymbol{F}$, there is a scalar function ϕ such that

$$\boldsymbol{F} = -\operatorname{grad} \phi$$

Example. Show that an electrostatic field is conservative by finding a function ϕ such that $\boldsymbol{E}(\rho) = -\operatorname{grad}(\phi)$ where $\boldsymbol{E}(\rho)$ is the resultant force at a point P at a distance ρ from a point charge e.

Now, in e.s.u. $$\boldsymbol{E}(\rho) = \frac{e}{\rho^2}\boldsymbol{\rho}$$

where $\boldsymbol{\rho}$ is the unit vector in the direction $\boldsymbol{OP}$. We define

$$\phi = e/\rho$$

Then
$$-\text{grad}\,\phi = -e\,\text{grad}\,(1/\rho)$$
$$= \frac{e}{\rho^2}\,\text{grad}\,\rho \qquad \text{using (iii) above.}$$

Now if the point charge is at the origin and P is the point (x, y, z)

$$\text{grad}\,\rho = \text{grad}\,(x^2+y^2+z^2)^{\frac{1}{2}}$$
$$= \frac{1}{2(x^2+y^2+z^2)^{\frac{1}{2}}}\,\text{grad}\,(x^2+y^2+z^2)$$
using (iii),
$$\text{grad}\,(x^2+y^2+z^2) = \boldsymbol{p}\frac{\partial}{\partial x}(x^2+y^2+z^2)+\boldsymbol{q}\frac{\partial}{\partial y}(x^2+y^2+z^2)$$
$$+\boldsymbol{r}\frac{\partial}{\partial z}(x^2+y^2+z^2)$$

therefore $$\text{grad}\,\rho = \frac{1}{2\rho}(2x\boldsymbol{p}+2y\boldsymbol{q}+2z\boldsymbol{r})$$

Now $$x\boldsymbol{p}+y\boldsymbol{q}+z\boldsymbol{r} = \rho\boldsymbol{\rho}$$

therefore $$\text{grad}\,\rho = \frac{1}{\rho}.\rho\boldsymbol{\rho}$$
$$= \boldsymbol{\rho}$$

therefore $$-\text{grad}\,\phi = \frac{e}{\rho^2}\boldsymbol{\rho} = \boldsymbol{E}(\rho)$$

Thus there is a function $\phi(= e/\rho)$ which satisfies the equation

$$-\text{grad}\,\phi = \boldsymbol{E}(\rho)$$

So the electrostatic field is conservative, and to find the work done in moving a particle from one point to another in an electrostatic field the actual path which the particle follows need not be considered. The function $\phi = e/\rho$ is the electrostatic scalar potential due to a point charge.

7.12 The divergence (div) of a vector field

We have considered in the previous section the operation of del on a scalar field. It may operate on a vector field $\boldsymbol{F}(x, y, z)$ in two ways.

Firstly, it can act on $\boldsymbol{F}$ as a scalar product, $\nabla . \boldsymbol{F}$, which is called the *divergence* of $\boldsymbol{F}$,

$$\operatorname{div} \boldsymbol{F} \equiv \nabla . \boldsymbol{F}$$

Secondly, it can act on $\boldsymbol{F}$ as a vector product, $\nabla \times \boldsymbol{F}$ which is called the *curl* of $\boldsymbol{F}$. This will be discussed later (7.14).

If F_x, F_y, F_z are the components of $\boldsymbol{F}$ and if the partial derivatives of these components exist, then

$$\begin{aligned}\operatorname{div} \boldsymbol{F} = \nabla . \boldsymbol{F} &= (\boldsymbol{p}(\partial/\partial x)+\boldsymbol{q}(\partial/\partial y)+\boldsymbol{r}(\partial/\partial z)).(F_x\boldsymbol{p}+F_y\boldsymbol{q}+F_z\boldsymbol{r}) \\ &= \partial F_x/\partial x+\partial F_y/\partial y+\partial F_z/\partial z\end{aligned}$$

Note that

$$\nabla . \boldsymbol{F} \neq \boldsymbol{F} . \nabla$$

$$\boldsymbol{F} . \nabla = F_x \partial/\partial x+F_y \partial/\partial y+F_z \partial/\partial z$$

i.e. $\boldsymbol{F}.\nabla$ is an operator, whereas $\nabla . \boldsymbol{F}$ is a scalar function. The physical significance of divergence may be appreciated by considering the flow of ith component, $\boldsymbol{J}_i$ into or out of a point under consideration.

If $\operatorname{div} \boldsymbol{J}_i > 0$ there is a net outflow of the ith component and the point is said to be a *source*.

If $\operatorname{div} \boldsymbol{J}_i < 0$ there is a net inflow of the ith component and the point is said to be a *sink*.

This leads us to the fact that the total flow Φ_i of the ith component out of a volume, V, is given by

$$\Phi_i = \int_{\substack{\text{whole}\\ \text{volume}}} \operatorname{div} \boldsymbol{J}_i \, \mathrm{d}V.$$

7.13 The Laplacian operator

If $\phi(x, y, z)$ represents a scalar field

$$\operatorname{grad} \phi = \nabla \phi = \boldsymbol{p}(\partial\phi/\partial x)+\boldsymbol{q}(\partial\phi/\partial y)+\boldsymbol{r}(\partial\phi/\partial z)$$

Now

$$\begin{aligned}\operatorname{div} \operatorname{grad} \phi = \nabla . \operatorname{grad} \phi &\equiv \nabla . \nabla \phi \equiv \nabla^2 \phi \\ &= (\partial/\partial x)(\partial\phi/\partial x)+(\partial/\partial y)(\partial\phi/\partial y)+(\partial/\partial z)(\partial\phi/\partial z) \\ &= (\partial^2\phi/\partial x^2)+(\partial^2\phi/\partial y^2)+(\partial^2\phi/\partial z^2)\end{aligned}$$

Thus, provided that it is operating on a scalar,

$$\nabla^2 = \text{div grad}$$

∇^2 is called the Laplacian operator.

When ∇^2 operates on a vector $\boldsymbol{F}(x, y, z)$ whose components F_x, F_y, F_z possess a second partial derivative, we have

$$\begin{aligned}\nabla^2 \boldsymbol{F} &= (\partial^2/\partial x^2 + \partial^2/\partial y^2 + \partial^2/\partial z^2)(F_x \boldsymbol{p} + F_y \boldsymbol{q} + F_z \boldsymbol{r})\\ &= \boldsymbol{p}\nabla^2 F_x + \boldsymbol{q}\nabla^2 F_y + \boldsymbol{r}\nabla^2 F_z\end{aligned}$$

Note that div grad $\boldsymbol{F}$ is not meaningful since grad is only defined when it operates on a scalar function.

Several important equations involve the Laplacian operator (see 4.6).

Laplace's equation $\qquad \nabla^2 \phi = 0$

Diffusion equation $\qquad \nabla^2 \phi = \dfrac{1}{a^2}\dfrac{\partial \phi}{\partial t}$

Wave equation $\qquad \nabla^2 \phi = \dfrac{1}{a^2}\dfrac{\partial^2 \phi}{\partial t^2}$

Schrödinger equation (time-independent)

$$\nabla^2 \phi + \frac{8\pi^2 m}{h^2}(E - U)\phi = 0$$

7.14 The curl of a vector field

If $\boldsymbol{F}(x, y, z)$ is a vector whose components F_x, F_y, F_z can be differentiated,

$$\begin{aligned}\text{curl } \boldsymbol{F} &\equiv \nabla \times \boldsymbol{F}\\ &= (\boldsymbol{p}(\partial/\partial x) + \boldsymbol{q}(\partial/\partial y) + \boldsymbol{r}(\partial/\partial z)) \times (F_x \boldsymbol{p} + F_y \boldsymbol{q} + F_z \boldsymbol{r})\\ &= \begin{vmatrix} \boldsymbol{p} & \boldsymbol{q} & \boldsymbol{r} \\ \partial/\partial x & \partial/\partial y & \partial/\partial z \\ F_x & F_y & F_z \end{vmatrix}\end{aligned}$$

i.e. curl $\boldsymbol{F}$

$$= \boldsymbol{p}\left(\frac{\partial F_z}{\partial y} - \frac{\partial F_y}{\partial z}\right) + \boldsymbol{q}\left(\frac{\partial F_x}{\partial z} - \frac{\partial F_z}{\partial x}\right) + \boldsymbol{r}\left(\frac{\partial F_y}{\partial x} - \frac{\partial F_x}{\partial y}\right) \quad \text{[see 8.2]}$$

The physical significance of the curl of a vector field is shown in the following example.

Example. A rigid body rotates with angular velocity ω about an axis through a fixed point O. The angular velocity of any point P in the body, $\boldsymbol{v}$, is given by $\boldsymbol{v} = \boldsymbol{\omega}\times\boldsymbol{\alpha}$ where $\boldsymbol{\alpha} = \boldsymbol{OP}$ and $\boldsymbol{\omega}$ is the angular velocity vector of the body. Show that curl $\boldsymbol{v} = 2\boldsymbol{\omega}$.

$$\boldsymbol{\omega}\times\boldsymbol{\alpha} = \begin{vmatrix} \boldsymbol{p} & \boldsymbol{q} & \boldsymbol{r} \\ \omega_x & \omega_y & \omega_z \\ x & y & z \end{vmatrix}$$

if $$\boldsymbol{\omega} = \omega_x\boldsymbol{p}+\omega_y\boldsymbol{q}+\omega_z\boldsymbol{r}$$

and $$\boldsymbol{\alpha} = x\boldsymbol{p}+y\boldsymbol{q}+z\boldsymbol{r}$$

i.e. $$\boldsymbol{\omega}\times\boldsymbol{\alpha} = \boldsymbol{p}(\omega_y z-\omega_z y)+\boldsymbol{q}(\omega_z x-\omega_x z)+\boldsymbol{r}(\omega_x y-\omega_y x)$$

therefore curl $\boldsymbol{v}$

$$= \nabla\times\boldsymbol{v} = \nabla\times(\boldsymbol{\omega}\times\boldsymbol{\alpha})$$

$$= \begin{vmatrix} \boldsymbol{p} & \boldsymbol{q} & \boldsymbol{r} \\ \partial/\partial x & \partial/\partial y & \partial/\partial z \\ \omega_y z-\omega_z y & \omega_z x-\omega_x z & \omega_x y-\omega_y x \end{vmatrix}$$

$$= \boldsymbol{p}[\partial/\partial y)(\omega_x y-\omega_y x)-(\partial/\partial z)(\omega_z x-\omega_x z)]$$

$$+\boldsymbol{q}[\partial/\partial z)(\omega_y z-\omega_z y)-(\partial/\partial x)(\omega_x y-\omega_y x)]$$

$$+\boldsymbol{r}[(\partial/\partial x)(\omega_z x-\omega_x z)-(\partial/\partial y)(\omega_y z-\omega_z y)]$$

$$= \boldsymbol{p}(\omega_x+\omega_x)+\boldsymbol{q}(\omega_y+\omega_y)+\boldsymbol{r}(\omega_z+\omega_z)$$

$$= 2\boldsymbol{\omega}$$

since ω_x, ω_y, ω_z are independent of x, y, z and $\partial x/\partial x = 1$ and $\partial y/\partial x = 0$.

Thus the curl of the velocity vector of any point is a function of the angular velocity vector of the body. In general curl is a measure of the rotational properties of a vector field.

7.15 Summary of useful formulae and theorems

$$\nabla = \boldsymbol{p}(\partial/\partial x)+\boldsymbol{q}(\partial/\partial y)+\boldsymbol{r}(\partial/\partial z)$$

$$\nabla^2 = (\partial^2/\partial x^2)+(\partial^2/\partial y^2)+(\partial^2/\partial z^2)$$

$$\text{grad }\phi = \nabla\phi = \boldsymbol{p}(\partial\phi/\partial x)+\boldsymbol{q}(\partial\phi/\partial y)+\boldsymbol{r}(\partial\phi/\partial z)$$

$$\text{div }\boldsymbol{F} = \nabla.\boldsymbol{F} = (\partial F_x/\partial y)+(\partial F_y/\partial y)+(\partial F_z/\partial z)$$

$$\nabla^2\phi = \nabla.\nabla\phi = (\partial^2\phi/\partial x^2)+(\partial^2\phi/\partial y^2)+(\partial^2\phi/\partial z^2)$$

$$\nabla^2 \boldsymbol{F} = \boldsymbol{p}\nabla^2 F_x+\boldsymbol{q}\nabla^2 F_y+\boldsymbol{r}\nabla^2 F_z$$

$$\text{curl}\, \boldsymbol{F} = \nabla\times\boldsymbol{F} = \begin{vmatrix} \boldsymbol{p} & \boldsymbol{q} & \boldsymbol{r} \\ \partial/\partial x & \partial/\partial y & \partial/\partial z \\ F_x & F_y & F_z \end{vmatrix}$$

$$\text{grad}\,(\phi\psi) = \phi\ \text{grad}\ \psi+\psi\ \text{grad}\ \phi$$

$$\text{div}\,(\phi\boldsymbol{F}) = \nabla(\phi\boldsymbol{F}) = \phi\nabla.\boldsymbol{F}+(\nabla.\phi).\boldsymbol{F}$$

$$\nabla.(\nabla\times\boldsymbol{F}) = 0$$

$$\text{div}\,(\boldsymbol{E}\times\boldsymbol{F}) = \nabla.(\boldsymbol{E}\times\boldsymbol{F}) = \boldsymbol{F}.(\nabla\times\boldsymbol{E})-\boldsymbol{E}.(\nabla\times\boldsymbol{F})$$

$$\text{curl}\,(\phi\boldsymbol{F}) = \nabla\times\phi\boldsymbol{F} = \phi\nabla\times\boldsymbol{F}+(\nabla\phi)\times\boldsymbol{F}$$

$$\nabla\times\nabla\phi = 0$$

$$\nabla\times(\nabla\times\boldsymbol{F}) = \nabla(\nabla.\boldsymbol{F})-\nabla^2\boldsymbol{F}$$

$$\text{div}\,(\boldsymbol{E}.\boldsymbol{F}) = \nabla.(\boldsymbol{E}.\boldsymbol{F}) = (\boldsymbol{E}.\nabla).\boldsymbol{F}+(\boldsymbol{F}.\nabla)\boldsymbol{E}+\boldsymbol{E}\times(\nabla\times\boldsymbol{F})+\boldsymbol{F}\times(\nabla\times\boldsymbol{E})$$

$$\text{curl}\,(\boldsymbol{E}\times\boldsymbol{F})$$

$$= \nabla\times(\boldsymbol{E}\times\boldsymbol{F}) = (\nabla.\boldsymbol{F}).\boldsymbol{E}-(\nabla.\boldsymbol{E}).\boldsymbol{F}+(\boldsymbol{F}.\nabla).\boldsymbol{E}-(\boldsymbol{E}.\nabla).\boldsymbol{F}$$

$$\boldsymbol{F}\times(\nabla\times\boldsymbol{F}) = \tfrac{1}{2}\nabla.\boldsymbol{F}^2-(\boldsymbol{F}.\nabla).\boldsymbol{F}$$

Gauss' theorem

If S is a closed surface enclosing a volume V and $\boldsymbol{F}$ is a continuous vector field whose components have partial derivatives, then

$$\int_S \boldsymbol{F}.\mathrm{d}\boldsymbol{S} = \int_V \text{div}\,\boldsymbol{F}.\mathrm{d}V$$

where $\mathrm{d}\boldsymbol{S}$ is a differential vector normal to the surface in an outward direction.

Stokes' theorem

If S is an open surface bounded by a closed curve C, and if $\boldsymbol{F}$ is a continuous vector field whose components have partial derivatives, then the line integral of $\boldsymbol{F}$

$$\oint_C \boldsymbol{F}\,\mathrm{d}\boldsymbol{\alpha} = \int_S \text{curl}\,\boldsymbol{F}.\mathrm{d}\boldsymbol{S}$$

where $\mathrm{d}\boldsymbol{\alpha}$ is a differential vector whose components are $\mathrm{d}x$, $\mathrm{d}y$, $\mathrm{d}z$ and $\mathrm{d}\boldsymbol{S}$ is normal to the surface in a direction derived by application of the right hand rule assuming that C is described in the positive direction.

Integral theorems

For a conservative field, $\boldsymbol{F}$

(i) the line integral from A to B, $\int_A^B \boldsymbol{F}.\mathrm{d}\boldsymbol{\alpha}$ is independent of the path from A to B within the region.

(ii) for every closed curve in the region $\oint_C \boldsymbol{F}.\mathrm{d}\boldsymbol{\alpha} = 0$.

(iii) there is a single valued function ϕ such that $\boldsymbol{F} = -\mathrm{grad}\ \phi$.

8 *Determinants and Matrices*

The use of matrices in quantum chemistry is widespread, but they are also of importance in other fields such as spectroscopy and statistical thermodynamics. In fact, in quantum mechanics, the matrix formulation was developed before Schrödinger's wave mechanics. The matrix formalism has the advantages (a) of enabling a simplification to be made in notation, and (b) in being especially suited to computer manipulation.

8.1 Matrices

A matrix is an array of mn quantities which are usually arranged in m rows and n columns. We may denote the matrix

$$\mathbf{A} = \begin{bmatrix} a_{11} & a_{12} & \dots & a_{1n} \\ \vdots & & & \\ a_{m1} & a_{m2} & \dots & a_{mn} \end{bmatrix}$$

and the elements of the matrix i.e. the individual quantities are labelled by two indices m and n where a_{mn} is the element in row m and column n.

One important and convenient use of matrix notation is in the abbreviation of sets of linear equations. For example the set of equations

$$\begin{aligned} a_{11}x_1 + a_{12}x_2 + \dots a_{1n}x_n &= y_1 \\ a_{21}x_1 + a_{22}x_2 + \dots a_{2n}x_n &= y_2 \\ \vdots \qquad\qquad & \quad \vdots \\ a_{m1}x_1 + a_{m2}x_2 + \dots a_{mn}x_n &= y_m \end{aligned}$$

may be written

$$\begin{bmatrix} a_{11} & a_{12} & \dots & a_{1n} \\ a_{21} & a_{22} & \dots & a_{2n} \\ \vdots & & & \\ a_{m1} & a_{m2} & \dots & a_{mn} \end{bmatrix} \begin{bmatrix} x_1 \\ x_2 \\ \vdots \\ x_n \end{bmatrix} = \begin{bmatrix} y_1 \\ y_2 \\ \vdots \\ y_m \end{bmatrix}$$

If $m = n$ then the matrix is known as a *square matrix* whereas if $m \neq n$ it is generally referred to as an $m \times n$ *matrix*. Some special cases may be considered. For example, $1 \times n$ matrices are normally known as *row matrices* and $m \times 1$ matrices as *column matrices*:

$$\begin{bmatrix} a_1 \\ a_2 \\ a_3 \\ a_4 \end{bmatrix} \text{ is a column matrix;}$$

$[a_1 a_2 a_3 a_4]$ is a row matrix.

In a square matrix the diagonal elements are a_{mm} and the sum of these is called the *trace* or the *character* of the matrix

e.g. the character of the matrix

$$\begin{bmatrix} 5 & -2 & 3 \\ 4 & 6 & 9 \\ 2 & 1 & 0 \end{bmatrix} \text{ is } 5+6+0 = 11$$

i.e.
$$\text{Tr}\,\mathbf{A} = \sum_m a_{mm}$$

Diagonal matrices have all the elements off the diagonal equal to zero. It is often necessary to reduce non-diagonal matrices to diagonal matrices for ease of manipulation, especially in quantum mechanics.

Scalar matrices are diagonal matrices with equal diagonal elements.

i.e.
$$\begin{bmatrix} a & 0 & 0 & 0 \\ 0 & a & 0 & 0 \\ 0 & 0 & a & 0 \\ 0 & 0 & 0 & a \end{bmatrix}$$

The *unit matrix* **1** is a scalar matrix with the diagonal elements equal to unity.

e.g.
$$\begin{bmatrix} 1 & 0 & 0 \\ 0 & 1 & 0 \\ 0 & 0 & 1 \end{bmatrix} \text{ or } \begin{bmatrix} 1 & 0 \\ 0 & 1 \end{bmatrix}$$

8.2 Determinants

If we have an $m \times n$ matrix $\mathbf{A}$, there may be associated with it a *determinant*

$$\det \mathbf{A} = \begin{vmatrix} a_{11} \dots a_{1n} \\ \vdots \\ a_{m1} \dots a_{mn} \end{vmatrix}$$

This is simply a function (real or complex) which may be associated with the elements of a matrix. If $\det \mathbf{A} = 0$, the matrix is *singular*; otherwise it is non-singular.

We only need at this stage to be familiar with some of the simplest properties of determinants.

1. If we interchange rows (or columns), the determinant changes sign.
2. If any row (or column) consists of only zeros, $\det \mathbf{A} = 0$.

Expansion of a Determinant

A determinant, unlike a matrix, may be expanded to give the value of the determinant. For example the second order determinant

$$\begin{vmatrix} a & b \\ c & d \end{vmatrix}$$

has the value $ad-bc$ and, in principle, all determinants of order n, $\det \mathbf{A}$, may be expanded as a total of $n!$ products. ($n! = n(n-1)(n-2) \dots 3.2.1$, see Appendix A4)

The expansion of third and higher order determinants may be done in terms of *cofactors*. Before defining cofactors we define the *minor* of the element a_{rs} in a determinant as the determinant of next lowest order formed by striking out row r and column s of the original determinant.

i.e.

$$\begin{vmatrix} a_{11} & a_{12} & a_{13} \\ a_{21} & a_{22} & a_{23} \\ a_{31} & a_{32} & a_{33} \end{vmatrix} \rightarrow \begin{vmatrix} a_{12} & a_{13} \\ a_{32} & a_{33} \end{vmatrix}$$

The minor of a_{21} is obtained above by striking out row 2 and column 1 as indicated. The minor of a_{13} obtained similarly:

$$\begin{vmatrix} a_{11} & a_{12} & a_{13} \\ a_{21} & a_{22} & a_{23} \\ a_{31} & a_{32} & a_{33} \end{vmatrix} \rightarrow \begin{vmatrix} a_{21} & a_{22} \\ a_{31} & a_{32} \end{vmatrix} \text{ minor of } a_{13}$$

The *cofactor* of a_{rs} is $(-1)^{r+s}$ times the *minor* of a_{rs}. The determinant is reduced to a sum of determinants of next lower order by taking the product of each element and its cofactor along the entire row. (The sign alternates from term to term.) i.e. expansion of the above 3rd order determinant gives

$$\det \mathbf{A} = \begin{vmatrix} a_{11} & a_{12} & a_{13} \\ a_{21} & a_{22} & a_{23} \\ a_{31} & a_{32} & a_{33} \end{vmatrix}$$

$$= \{(-1)^{1+1}a_{11}\}\begin{vmatrix} a_{22} & a_{23} \\ a_{32} & a_{33} \end{vmatrix} + \{(-1)^{1+2}a_{12}\}\begin{vmatrix} a_{21} & a_{23} \\ a_{31} & a_{33} \end{vmatrix}$$

$$+\{(-1)^{1+3}a_{13}\}\begin{vmatrix} a_{21} & a_{22} \\ a_{31} & a_{32} \end{vmatrix}$$

$$= a_{11}\begin{vmatrix} a_{22} & a_{23} \\ a_{32} & a_{33} \end{vmatrix} - a_{12}\begin{vmatrix} a_{21} & a_{23} \\ a_{31} & a_{33} \end{vmatrix} + a_{13}\begin{vmatrix} a_{21} & a_{22} \\ a_{31} & a_{32} \end{vmatrix}$$

The lower order determinants (here 2nd order) are expanded in the usual way:

$$\begin{vmatrix} a_{22} & a_{23} \\ a_{32} & a_{33} \end{vmatrix} = a_{22}.a_{33} - a_{32}.a_{23}$$

giving the final result:

$$\det \mathbf{A} = a_{11}a_{22}a_{33} - a_{11}a_{23}a_{32} - a_{12}a_{21}a_{33}$$
$$+ a_{12}a_{23}a_{31} + a_{13}a_{21}a_{32} - a_{13}a_{22}a_{31}$$

Expansion of higher order determinants is best carried out by this method. As an example of an expansion we have

$$\begin{vmatrix} 1 & -1 & 2 \\ 0 & 3 & 1 \\ 4 & 2 & -1 \end{vmatrix} = 1\begin{vmatrix} 3 & 1 \\ 2 & -1 \end{vmatrix} - (-1)\begin{vmatrix} 0 & 1 \\ 4 & -1 \end{vmatrix} + 2\begin{vmatrix} 0 & 3 \\ 4 & 2 \end{vmatrix}$$

$$= 1\{3(-1)-2(1)\}+1\{0(-1)-1(4)\}+2\{0(2)-3(4)\}$$
$$= 1(-5)+1(-4)+2(-12)$$
$$= -5-4-24$$
$$= -33$$

8.3 Operations with matrices

(1) *Addition of matrices*

If **A** and **B** are matrices of the same dimensions then

S = **A**+**B** is defined by its matrix elements s_{mn} which are

$$s_{mn} = a_{mn}+b_{mn}$$

e.g.
$$\begin{bmatrix} 3 & 2 & 0 \\ -1 & 0 & 2 \end{bmatrix} + \begin{bmatrix} 6 & -2 & 9 \\ 5 & 4 & 0 \end{bmatrix} = \begin{bmatrix} 9 & 0 & 9 \\ 4 & 4 & 2 \end{bmatrix}$$

(2) *Multiplication by a constant* (*scalar multiplication*)

When **A** is multiplied by a number K then the elements of the matrix $\mathbf{P}(=K\mathbf{A})$ are Ka_{mn}

e.g.
$$3\times\begin{bmatrix} 1 & 2 \\ -3 & 0 \\ 5 & 7 \end{bmatrix} = \begin{bmatrix} 3 & 6 \\ -9 & 0 \\ 15 & 21 \end{bmatrix}$$

(3) *Multiplication of two matrices*

Two matrices **A** and **B** may only be multiplied to give a product

$$\mathbf{AB} = \mathbf{C}$$

if the number of *columns* in **A** is equal to the number of *rows* in **B**. If **A** is of order $m\times n$ and **B** is of order $n\times p$ then the product **C** exists and is a matrix of order $m\times p$.

e.g.
$$\begin{bmatrix} a_{11} & a_{12} \\ a_{21} & a_{22} \\ a_{31} & a_{32} \\ a_{41} & a_{42} \end{bmatrix} \begin{bmatrix} b_{11} & b_{12} & b_{13} \\ b_{21} & b_{22} & b_{23} \end{bmatrix} = \begin{bmatrix} c_{11} & c_{12} & c_{13} \\ c_{21} & c_{22} & c_{23} \\ c_{31} & c_{32} & c_{33} \\ c_{41} & c_{42} & c_{43} \end{bmatrix}$$

4×2 matrix 2×3 matrix 4×3 matrix

N.B. Multiplication can only be carried out under the above conditions, i.e. only if the matrices are *conformable*.

The elements of matrix **C** may be calculated from the expression:

$$c_{ij} = \sum_{k=1}^{n} a_{ik}b_{kj}.$$

Example. Find the product **AB** where

$$\mathbf{A} = \begin{bmatrix} 1 & -2 & 3 \\ -4 & 2 & 5 \end{bmatrix} \text{ and } \mathbf{B} = \begin{bmatrix} 1 & 3 \\ -1 & 0 \\ 2 & 4 \end{bmatrix}$$

Let

$$\begin{bmatrix} 1 & -2 & 3 \\ -4 & 2 & 5 \end{bmatrix} \begin{bmatrix} 1 & 3 \\ -1 & 0 \\ 2 & 4 \end{bmatrix} = \begin{bmatrix} c_{11} & c_{12} \\ c_{21} & c_{22} \end{bmatrix}$$

Then

$$\begin{aligned} c_{11} &= \sum_{k=1}^{3} a_{1k}b_{k1} \\ &= a_{11}b_{11}+a_{12}b_{21}+a_{13}b_{31} \\ &= (1)(1)+(-2)(-1)+(3)(2) \\ &= 9 \end{aligned}$$

$$\begin{aligned} c_{12} &= \sum_{k=1}^{3} a_{1k}b_{k2} \\ &= a_{11}b_{12}+a_{12}b_{22}+a_{13}b_{32} \\ &= (1)(3)+(-2)(0)+(3)(4) \\ &= 15. \end{aligned}$$

Similarly $c_{21} = 4$

and $c_{22} = 8.$

Hence

$$\mathbf{AB} = \begin{bmatrix} c_{11} & c_{12} \\ c_{21} & c_{22} \end{bmatrix} = \begin{bmatrix} 9 & 15 \\ 4 & 8 \end{bmatrix}$$

8.4 Special matrices

One can carry out a number of operations, or changes, in the elements a_{mn} of A and thus obtain a series of new matrices. Several of these frequently occur in quantum mechanical problems and these will be defined here.

(1) *The transpose of* **A**; $\tilde{\mathbf{A}}$

The transpose of **A** is another matrix in which the rows and columns are interchanged. i.e. the element a_{mn} in **A** becomes a_{nm} in $\tilde{\mathbf{A}}$. Therefore if the dimensions of **A** are $m \times n$ the dimensions of $\tilde{\mathbf{A}}$ are $n \times m$. In particular the transpose of a row matrix is a column matrix and vice versa.

e.g. if $\mathbf{A} = \begin{bmatrix} 5 & 4 & 2 \\ 3 & 1 & -1 \end{bmatrix}$ then $\tilde{\mathbf{A}} = \begin{bmatrix} 5 & 3 \\ 4 & 1 \\ 2 & -1 \end{bmatrix}$

(2) *The complex conjugate of* **A**; **A***

The complex conjugate of **A** is obtained by substituting the complex conjugate of each element in **A**.

e.g.

$$\text{if } \mathbf{A} = \begin{bmatrix} 3 & -3i \\ i & -2 \\ 1+i & 4 \end{bmatrix} \text{ then } \mathbf{A}^* = \begin{bmatrix} 3 & 3i \\ -i & -2 \\ 1-i & 4 \end{bmatrix}$$

(3) *The adjoint matrix of* **A**; *adj.* **A** *or* $\hat{\mathbf{A}}$

The adjoint matrix of **A** is obtained by replacing each element of **A** by its cofactor and then transposing.

e.g.

$$\text{if } \mathbf{A} = \begin{bmatrix} 1 & 2 & 3 \\ 1 & 3 & 5 \\ 1 & 5 & 12 \end{bmatrix}$$

replacing each element by its cofactor (see 8.2) gives

$$\begin{bmatrix} 11 & -7 & 2 \\ -9 & 9 & -3 \\ 1 & -2 & 1 \end{bmatrix}$$

and transposing this matrix gives

$$\text{adj. } \mathbf{A} = \begin{bmatrix} 11 & -9 & 1 \\ -7 & 9 & -2 \\ 2 & -3 & 1 \end{bmatrix}$$

In quantum mechanics the term adjoint matrix is sometimes used to refer to the Hermitian conjugate which is discussed below.

(4) *The inverse matrix of* $\mathbf{A}$; $\mathbf{A}^{-1}$

The inverse or reciprocal matrix of $\mathbf{A}$ is that matrix $\mathbf{A}^{-1}$ which satisfies

$$\mathbf{A}\mathbf{A}^{-1} = \mathbf{1} \qquad \text{(where } \mathbf{1} \text{ is the unit matrix)}$$

The inverse matrix may be determined from the identity

$$\mathbf{A}^{-1} = \frac{\text{adj. } \mathbf{A}}{|\mathbf{A}|}$$

e.g.

$$\text{if} \quad \mathbf{A} = \begin{bmatrix} 1 & 2 & 3 \\ 1 & 3 & 5 \\ 1 & 5 & 12 \end{bmatrix} \quad \text{as in the previous example,}$$

then we know

$$\text{adj. } \mathbf{A} = \begin{bmatrix} 11 & -9 & 1 \\ -7 & 9 & -2 \\ 2 & -3 & 1 \end{bmatrix}$$

$$|\mathbf{A}| = \det \mathbf{A} = \begin{vmatrix} 1 & 2 & 3 \\ 1 & 3 & 5 \\ 1 & 5 & 12 \end{vmatrix}$$

$$= 1(11) - 2(7) + 3(2)$$

$$= 3$$

therefore

$$\mathbf{A}^{-1} = \begin{bmatrix} \frac{11}{3} & -3 & \frac{1}{3} \\ -\frac{7}{3} & 3 & -\frac{2}{3} \\ \frac{2}{3} & -1 & \frac{1}{3} \end{bmatrix}$$

One may check this by showing

$$\begin{bmatrix} 1 & 2 & 3 \\ 1 & 3 & 5 \\ 1 & 5 & 12 \end{bmatrix} \begin{bmatrix} \frac{11}{3} & -3 & \frac{1}{3} \\ -\frac{7}{3} & 3 & -\frac{2}{3} \\ \frac{2}{3} & -1 & \frac{1}{3} \end{bmatrix} = \begin{bmatrix} 1 & 0 & 0 \\ 0 & 1 & 0 \\ 0 & 0 & 1 \end{bmatrix} = \mathbf{1}$$

(5) *The Hermitian conjugate of* $\mathbf{A}$; $\mathbf{A}^{\dagger}$

The Hermitian conjugate of $\mathbf{A}$ is obtained by transposing the complex conjugate matrix.

e.g.

$$\text{if}\quad \mathbf{A} = \begin{bmatrix} 3 & -3i \\ i & -2 \\ 1+i & 4 \end{bmatrix}$$

$$\mathbf{A}^* = \begin{bmatrix} 3 & 3i \\ -i & -2 \\ 1-i & 4 \end{bmatrix}$$

and thus

$$\mathbf{A}^\dagger = \begin{bmatrix} 3 & -i & 1-i \\ 3i & -2 & 4 \end{bmatrix}$$

(6) *Matrices which obey particular equations*

(a) *Orthogonal matrices* are matrices whose inverse and transpose are identical:

$$\mathbf{A}^{-1} = \tilde{\mathbf{A}}$$

i.e.

$$\mathbf{A}\tilde{\mathbf{A}} = \tilde{\mathbf{A}}\mathbf{A} = \mathbf{1}$$

(b) *Hermitian matrices* are matrices identical to their Hermitian conjugates:

$$\mathbf{A} = \mathbf{A}^\dagger$$

(c) *Unitary matrices* are matrices whose inverse matrices are identical to their Hermitian conjugates:

$$\mathbf{A}^{-1} = \mathbf{A}^\dagger$$

8.5 Linear transformations

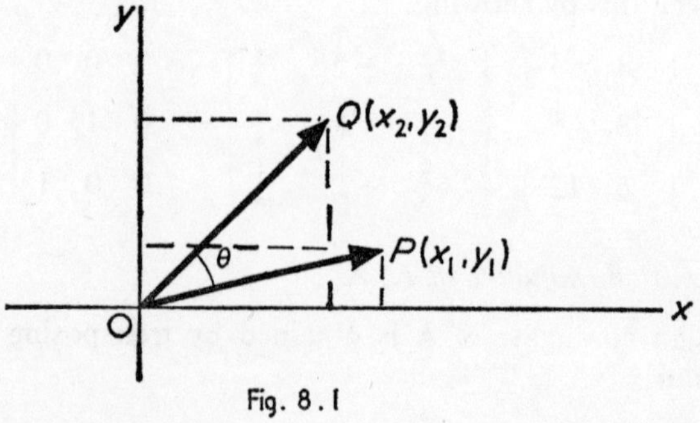

Fig. 8.1

If the vector $\boldsymbol{OP}$ is rotated counter-clockwise through an angle θ (Fig. 8.1), its components are changed from (x_1, y_1) to (x_2, y_2). These two sets of components are related by the equations,

$$x_2 = x_1 \cos\theta - y_1 \sin\theta,$$

$$y_2 = x_1 \sin\theta + y_1 \cos\theta,$$

as may be shown by trigonometry. Now the components of a vector may be written in the form of a column matrix. Thus $\boldsymbol{OP}$ before transformation (rotation in this case) may be written as $\begin{bmatrix} x_1 \\ y_1 \end{bmatrix}$ and after transformation as $\begin{bmatrix} x_2 \\ y_2 \end{bmatrix}$. In other words, the rotation has *transformed* the matrix $\begin{bmatrix} x_1 \\ y_1 \end{bmatrix}$ into a new matrix $\begin{bmatrix} x_2 \\ y_2 \end{bmatrix}$. If we can find a matrix $\mathbf{T}$ which will obey the relationship

$$\begin{bmatrix} x_2 \\ y_2 \end{bmatrix} = \mathbf{T} \begin{bmatrix} x_1 \\ y_1 \end{bmatrix}$$

then $\mathbf{T}$ describes the transformation of the vector $\boldsymbol{OP}$, and is known as the *transformation matrix*. It will in this particular case have to be a 2×2 matrix to obey the conditions of matrix multiplication (8.3).

From the equations relating the components given above, $\mathbf{T}$ in this case is the matrix

$$\begin{bmatrix} \cos\theta & -\sin\theta \\ \sin\theta & \cos\theta \end{bmatrix}$$

An n-dimensional vector may be written as a column matrix

$$\begin{bmatrix} a_1 \\ a_2 \\ \vdots \\ a_n \end{bmatrix}$$

and may be similarly transformed by an $n \times n$ matrix

$$\begin{bmatrix} b_1 \\ b_2 \\ \vdots \\ b_n \end{bmatrix} = \begin{bmatrix} c_{11} & c_{12} \cdots c_{1n} \\ c_{12} & & \\ \vdots & & \\ c_{n1} & & c_{nn} \end{bmatrix} \begin{bmatrix} a_1 \\ a_2 \\ \vdots \\ a_n \end{bmatrix}$$

Transformation of any vector by the corresponding unit matrix is equivalent to leaving the vector unchanged.

e.g.
$$\begin{bmatrix} x_1 \\ y_1 \end{bmatrix} = \begin{bmatrix} 1 & 0 \\ 0 & 1 \end{bmatrix} \begin{bmatrix} x_1 \\ y_1 \end{bmatrix}$$

8.6 Scalar products in matrix form

In chapter 7 it was shown that the scalar product of two vectors is a number. If we write the vectors in matrix form

$$\text{say,} \quad \mathbf{A} = \begin{bmatrix} a_1 \\ a_2 \\ \vdots \\ a_n \end{bmatrix} \quad \text{and} \quad \mathbf{B} = \begin{bmatrix} b_1 \\ b_2 \\ \vdots \\ b_n \end{bmatrix}$$

then it is impossible to multiply **A** by **B**, but it is possible to multiply **A** by the transpose of **B**, $\tilde{\mathbf{B}}$ which is a row vector.

$$\mathbf{A}\tilde{\mathbf{B}} = \begin{bmatrix} a_1 \\ a_2 \\ \vdots \\ a_n \end{bmatrix} [b_1 \quad b_2 \ldots b_n]$$
$$= a_1 b_1 + a_2 b_2 + \ldots + a_n b_n$$
$$= \text{a number.}$$

$\tilde{\mathbf{B}}$ is identical to the scalar product previously derived in chapter 7.

i.e.
$$\mathbf{A}\tilde{\mathbf{B}} \equiv A.B$$

8.7 Matrices and the eigenvalue problem

In the transformation
$$\mathbf{AB} = \mathbf{C}$$

the matrix **A** acts on the vector represented by **B** turning it into vector represented by **C**. In 8.5 the vector ***OP*** was changed in direction but not in magnitude since it only involved rotation of the vector. It is also possible to transform **B** into a vector where both the magnitude and the direction have changed.

If on the other hand **A** is a scalar matrix, then only the vector magnitude alters and the direction remains the same.

i.e.
$$\mathbf{AB} = \lambda \mathbf{B}$$

where λ is a number. In this case **B** is the *eigenvector* of **A** and λ is its *eigenvalue*. (You may recognize this as being similar to the Schrödinger equation, $\mathscr{H}\Psi = E\Psi$.)

$$\text{If } \mathbf{A} = \begin{bmatrix} a_{11} & a_{12} \dots a_{1n} \\ \vdots & \\ a_{n1} & a_{nn} \end{bmatrix}$$

$$\text{and } \mathbf{B} = \begin{bmatrix} b_1 \\ b_2 \\ \vdots \\ b_n \end{bmatrix}$$ ... a column matrix since it represents a vector,

then writing $\mathbf{AB} = \lambda\mathbf{B}$

gives

$$\begin{bmatrix} a_{11} & a_{12} \dots a_{1n} \\ \vdots & \\ \vdots & \\ a_{n1} & a_{nn} \end{bmatrix} \begin{bmatrix} b_1 \\ b_2 \\ \vdots \\ b_n \end{bmatrix} = \lambda \begin{bmatrix} b_1 \\ b_2 \\ \vdots \\ b_n \end{bmatrix}$$

therefore

$$\begin{bmatrix} a_{11}b_1 & a_{12}b_2 \dots a_{1n}b_n \\ \vdots & \\ \vdots & \\ a_{n1}b_1 & a_{nn}b_n \end{bmatrix} = \begin{bmatrix} \lambda b_1 \\ \lambda b_2 \\ \vdots \\ \lambda b_n \end{bmatrix}$$

This is equivalent to the set of n linear equations

$$\begin{aligned} a_{11}b_1 + a_{12}b_2 + \dots + a_{1n}b_n &= \lambda b_1 \\ \vdots & \\ a_{n1}b_1 + \dots\dots\dots\dots + a_{nn}b_n &= \lambda b_n \end{aligned}$$

or

$$\begin{aligned} (a_{11}-\lambda)b_1 + a_{12}b_2 + \dots + a_{1n}b_n &= 0 \\ \vdots & \\ a_{n1}b_1 + a_{n2}b_2 + \dots\dots\dots + (a_{nn}-\lambda)b_n &= 0 \end{aligned}$$

These are known as the *secular equations* and there are n solutions for the value of λ.

The eigenvalue can be written as

$$(\mathbf{A}-\lambda\mathbf{1})\mathbf{B} = 0$$

and for non-zero **B** it can be shown that

$$|\mathbf{A}-\lambda\mathbf{1}| = 0$$

This determinant when expanded gives an nth order polynomial in λ, known as the *characteristic polynomial* which has n solutions of λ, the eigenvalues.

9 *Expansions of Functions*

9.1 Power series

A power series may be written

$$a_0+a_1x+a_2x^2+a_3x^3+\dots$$

or in more concise notation as $\sum_{n=0}^{\infty} a_nx^n$. The coefficients a_n are constants and all the terms are integral powers of $x(a_0 \equiv a_0x^0)$. Many functions may be written as power series and the series may then be used to evaluate the function. For example, sin x and cos x may be written as the power series,

$$\sin x = x-\frac{x^3}{3!}+\frac{x^5}{5!}-\frac{x^7}{7!}+\dots$$

$$\cos x = 1-\frac{x^2}{2!}+\frac{x^4}{4!}-\frac{x^6}{6!}+\dots$$

Exponential functions may also be represented by a power series,

$$e^x = 1+\frac{x}{1!}+\frac{x^2}{2!}+\frac{x^3}{3!}+\dots$$

A power series consists therefore of a set of functions x^n, each with its own coefficient which may be zero: e.g. in the power series for sin x all coefficients $a_n = 0$ when n is even, and for cos x, $a_n = 0$ when n is odd. This set of functions is known as the *basis set* and will only be complete if f(x) may be completely expressed by it. The three functions above, sin x, cos x and e^x each have an infinite set of basis functions. Any evaluation of e^x, for example, without using an infinite number of terms, which is certainly impracticable, will involve using an incomplete basis set. If we use an incomplete basis set, say up to x^7, we obtain the following value for e (i.e. e^1).

$$e = 1+\frac{1}{1!}+\frac{1}{2!}+\frac{1}{3!}+\frac{1}{4!}+\frac{1}{5!}+\frac{1}{6!}+\frac{1}{7!}$$

$$= 1+1+0{\cdot}5+0{\cdot}16167+0{\cdot}04167+0{\cdot}00833+0{\cdot}00138+0{\cdot}00020$$

$$= 2{\cdot}71325$$

which is sufficiently accurate for most purposes. If a value of e accurate to only one decimal place is satisfactory then a smaller basis set, up to x^3, would be sufficient.

It is thus often possible to evaluate functions using an incomplete set of basis functions and obtain answers of sufficient accuracy even when the series is infinite. The size of the incomplete set of functions which one has to use will depend on three factors,

(a) the nature of the function,

(b) the value of the variable (e.g. e^2 *converges* much more slowly than e: accuracy to one decimal place requires a basis set of up to x^6), and

(c) the accuracy to which the function is to be determined.

Some functions may be exactly described by an incomplete set of basis functions. For example, the function $(1+x^2)^3$ is exactly described by the set x^0, x^1, x^2, x^3, x^4, x^5 and x^6.

The power series associated with a well behaved function, $\mathrm{f}(x)$, (i.e. a function with no discontinuities) can be derived using the *MacLaurin series*,

$$\mathrm{f}(x) = \mathrm{f}(0)+\frac{x}{1!}\mathrm{f}'(0)+\frac{x^2}{2!}\mathrm{f}''(0)+\frac{x^3}{3!}\mathrm{f}'''(0)+\ldots$$

(The MacLaurin series is a special case of the more general *Taylor series*

$$\mathrm{f}(x) = \mathrm{f}(a)+\frac{(x-a)}{1!}\mathrm{f}'(a)+\frac{(x-a)^2}{2!}\mathrm{f}''(a)+\frac{(x-a)^3}{3!}\mathrm{f}'''(a)+\ldots$$

with $a = 0$.)

Take for example $\mathrm{f}(x) = \sin x$

$$\mathrm{f}(x) = \sin x \qquad \mathrm{f}(0) = \sin 0 = 0$$

$$\mathrm{f}'(x) = \cos x \qquad \frac{x}{1!}\mathrm{f}'(0) = x\cos 0 = x$$

$$\mathrm{f}''(x) = -\sin x \qquad \frac{x^2}{2!}\mathrm{f}''(0) = 0$$

$$\mathrm{f}'''(x) = -\cos x \qquad \frac{x^3}{3!}\mathrm{f}'''(0) = -\frac{x^3}{3!}, \quad \text{etc.}$$

and thus $\mathrm{f}(x) = \sin x = x-\frac{x^3}{3!}+\ldots$

Cos x may be expanded similarly or if the series for sin x is known the power series for cos x may be obtained by differentiation since

$$\frac{d}{dx}(\sin x) = \cos x$$

Two other power series are of fairly general application, namely,

$$\ln(1+x) = x - \frac{x^2}{2} + \frac{x^3}{3} - \frac{x^4}{4} + \ldots \qquad \text{for } -1 < x \leqq 1$$

and the binomial expansion

$$(1+x)^p = 1 + px + \frac{p(p-1)}{2!}x^2 + \frac{p(p-1)(p-2)}{3!}x^3 + \ldots \qquad \text{for } |x| < 1$$

These two series, unlike those for sin x, cos x and e^x, are only convergent (i.e. they have a limiting value) for certain values of x. The binomial expansion is an infinite series if p is non-integral or less than zero; otherwise it is finite.

Power series are often useful in computing functions for which tables are either not available or are not of sufficient accuracy. However, to do this efficiently the series should converge rapidly and then only a small basis set of functions (even if incomplete) will be sufficient.

Not all functions can be expanded as power series. For instance a power series becomes equal to a_0 when $x = 0$ and therefore any function which is infinite at $x = 0$ such as $1/x$ or ln x cannot be represented by such a series. Similarly a discontinuous function cannot be expanded as a power series. It is sometimes more convenient to express functions in terms of a *Fourier series* rather than a power series and this is discussed in the following section.

9.2 Fourier series

Fourier series are expansions in terms of sines and cosines rather than the x^n of the power series, and are very useful for describing periodic functions, from the motion of the simple pendulum as one example to the more complicated three dimensional diffraction of X-rays by a crystal. The general form of a Fourier series may be written

$$\begin{aligned} f(x) = {} & a_0 + a_1 \cos x + a_2 \cos 2x + \ldots \\ & + b_1 \sin x + b_2 \sin 2x + \ldots \end{aligned}$$

or as

$$f(x) = a_0 + \sum_{n=1}^{\infty} (a_n \cos nx + b_n \sin nx)$$

There are two sets of coefficients a_n and b_n. x takes values between π and $-\pi$.

Sines and cosines are periodic functions and repeat over a period of 2π (Fig. 9.1).

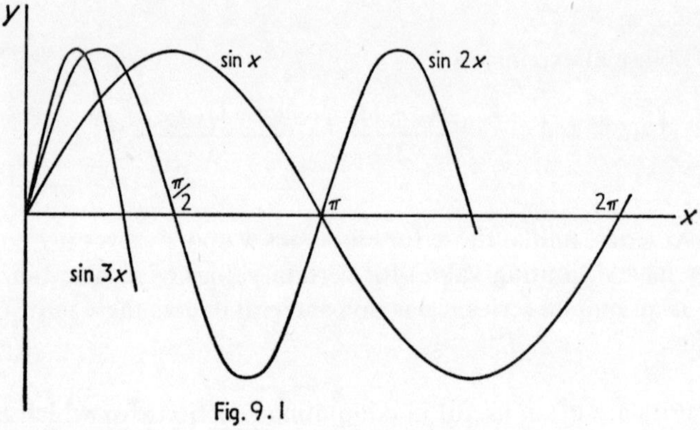

Fig. 9.1

Sin $2x$ and sin $3x$ may be regarded as the first and second harmonics of sin x. Similarly cos nx is a harmonic of cos x.

Any periodic function or waveform (Fig. 9.2) can be described by a Fourier series as a sum of harmonics on a particular fundamental by suitable choice of the coefficients a_n and b_n.

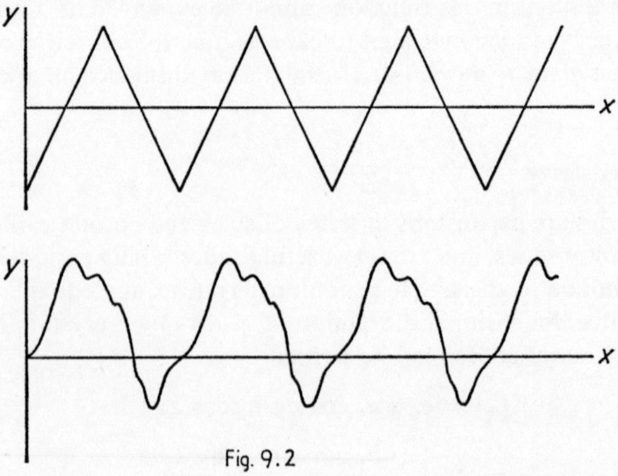

Fig. 9.2

The Fourier series may also be used to describe discontinuous functions (Fig. 9.3) which cannot be described by power series:

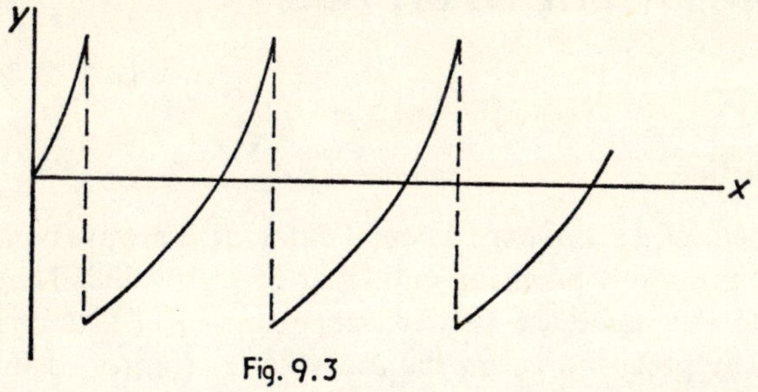

Fig. 9.3

Fourier series may be expressed in terms of complex numbers.

Since $$\sin(nx) = \frac{e^{inx} - e^{-inx}}{2i}$$

and $$\cos(nx) = \frac{e^{inx} + e^{-inx}}{2}$$

$$\begin{aligned} f(x) &= c_0 + c_1 e^{ix} + c_{-1} e^{-ix} + c_2 e^{2ix} + c_{-2} e^{-2ix} + \dots \\ &= \sum_{n=-\infty}^{\infty} c_n e^{inx} \end{aligned}$$

Methods of determining the coefficients of a Fourier series are given in standard mathematical texts. These series have very wide application in optics, acoustics, electronics, in fact in any subject concerned with wave motion. Two of the more important applications in chemistry occur in X-ray crystallography and in the interpretation of very low frequency infra-red spectroscopic data.

10 *Statistical Methods*

10.1 Introduction

In the treatment of a set of experimental data one is always concerned with assessing the accuracy, precision and error of the results. The accuracy is dependent on the difference between an observed measurement and the true value. The precision is, on the other hand, controlled by the differences found within a set of observed measurements. It is possible to obtain very precise results which are not accurate (e.g. by careful operation of incorrectly calibrated apparatus) but it is not possible to attain high accuracy without precision. Two different types of error may arise in scientific measurements, namely,

(a) determinate error, and

(b) indeterminate (or random) error.

A determinate error is one which affects each measurement in a set of measurements equally and such error will be reproducible—this type of error would account for the precise but inaccurate results mentioned above. It may however be minimised by careful experimental technique. Random errors are not reproducible and cannot be completely eliminated. In order to deal with random error, statistical methods are used. In practice it is often difficult to completely distinguish determinate and random error before some form of statistical analysis has been undertaken.

Given a set of experimental results $x_1, x_2, \ldots x_n$ one of the simplest computations to do is to find the *average* or *arithmetical mean*, $\bar{x}$, where

$$\bar{x} = \frac{x_1 + x_2 + \ldots x_n}{n}$$

However, the arithmetic mean alone gives no indication of the spread of the results. For example, the two sets of results below give the same values of $\bar{x}$ but one would intuitively have more confidence in the second set.

Set 1. 9·1, 10·5, 12·5, 7·6, 8·2, 11·8, 10·6, 9·7.

Set 2. 9·99, 10·00, 10·03, 9·99, 10·00, 9·98, 10·00, 10·01.

To obtain a clearer picture of the spread or frequency distribution of a set of results the data may be represented by a *histogram*. This plots on the

y-axis the number of times a given observation occurs (i.e. the frequency) against on the x-axis the range of observed values. A typical example is given below where a large number of measurements of the melting point of an alloy are represented by a histogram (Fig. 10.1).

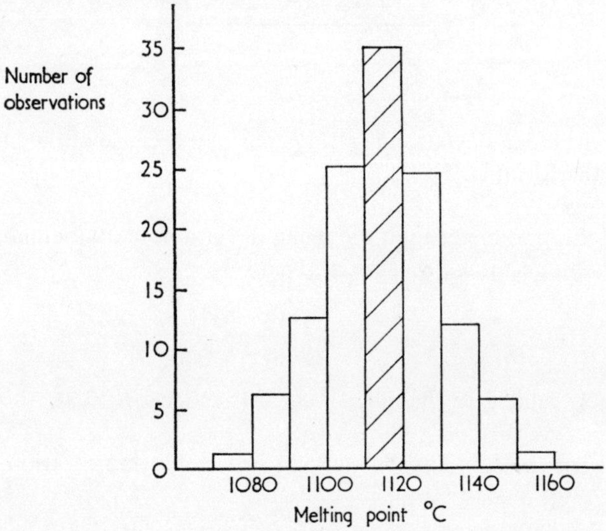

Fig. 10.1

The hatched area indicates that 35 of the measurements gave values of between 1110 and 1120°C for the melting point. The histogram is almost symmetrical indicating that there is an almost equal chance of a given observation being above or below the arithmetic mean which is 1115°. Also the probability of obtaining a value close to the mean is much higher than obtaining one far away from the mean. Although not all frequency distributions are symmetrical we shall leave discussion of the asymmetrical types until later in this chapter.

10.2 Mean deviation, variance and standard deviation

The use of a histogram to represent a set of results, although useful to indicate the symmetry of spread, is a little cumbersome for general usage. The spread or dispersion of the results may be given in terms of the *mean deviation* of n results, $x_1, x_2, \ldots x_n$.

This is given by

$$\text{Mean deviation} = \frac{\Sigma|x_i - \bar{x}|}{n}$$

$|x_i - \bar{x}|$ is the *modulus* of $x_i - \bar{x}$.

Example. Find the mean deviation of the five results 8·5, 9·5, 10, 10·5, 11·5.

$$\bar{x} = \frac{8{\cdot}5+9{\cdot}5+10+10{\cdot}5+11{\cdot}5}{5} = 10{\cdot}0$$

$$\frac{\Sigma|x_i-\bar{x}|}{n} = \frac{1{\cdot}5+0{\cdot}5+0{\cdot}0+0{\cdot}5+1{\cdot}5}{5} = \frac{4{\cdot}0}{5}$$

$$= 0{\cdot}8.$$

Note that the mean deviation is always positive because of the inclusion of $|x_i-\bar{x}|$ in the formula.

A more useful parameter than the mean deviation, in mathematical terms, is the *standard deviation*, σ. This is given by

$$\sigma = \sqrt{\frac{\Sigma(x-\bar{x})^2}{n-1}}$$

and is closely related to the *normal curve* (see Section 10.3).

Example. Find the standard deviation of the following titration figures: 4·95, 5·00, 5·11, 5·04, 4·90 ml.

$$\bar{x} = 5{\cdot}00 \qquad n = 5$$

x	$(x-\bar{x})$	$(x-\bar{x})^2$
4·95	−0·05	0·0025
5·00	0·00	0·0000
5·11	0·11	0·0121
5·04	0·04	0·0016
4·90	−0·10	0·0100

$$\Sigma(x-\bar{x})^2 = 0{\cdot}0262$$

$$\frac{\Sigma(x-\bar{x})^2}{n-1} = \frac{0{\cdot}0262}{4} = 0{\cdot}0066$$

therefore $\sigma = \sqrt{0{\cdot}0066} = 0{\cdot}081$ ml.

Note that the standard deviation and the mean deviation are in the same units as the results.

10.3 The normal curve

If sufficient readings are taken of any observable where there is only random error a frequency diagram of the type shown in Fig. 10.2 may be drawn.

This is only the limiting case of a histogram where the regions of measurement have become infinitely small.

In 1733 de Moivre found that such a curve, the *normal curve*, could be described by the formula,

$$f_{(x)} = \frac{1}{\sigma\sqrt{(2\pi)}} \exp\left(-\tfrac{1}{2}\left(\frac{x-\bar{x}}{\sigma}\right)^2\right)$$

where $f_{(x)}$ = frequency of occurrence of any given value,

$\bar{x}$ = arithmetic mean, and

σ = standard deviation.

The entire distribution is thus described by the mean and the standard deviation. The importance of the standard deviation is now obvious.

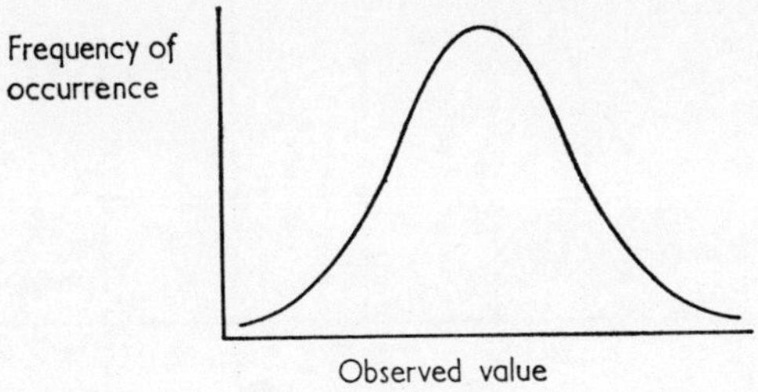

Fig. 10.2

10.4 Probability of random error

The normal curve allows the prediction of the probability of error. Probability, p, may be defined as the fraction of the number of times a particular result occurs in a large number of equally likely and mutually exclusive results. p can therefore have any value $0 \leqq p \leqq 1$. For example, the probability of obtaining heads from a tossed coin is 0·50 and a 6 from a die is 0·167.

Example. What is the probability of throwing either a 5 or a 6 with a die?

Total number of possible results = 6

Possible ways of obtaining desired result (i.e. a 5 or a 6) = 2

Probability $= \frac{2}{6} = 0{\cdot}33$

One can also calculate the probability, $p(n_1, n_2)$, of two specific events n_1 and n_2 occurring simultaneously from

$$p(n_1, n_2) = p(n_1).p(n_2)$$

or for several events, $n_1, n_2, \ldots n_n$

$$p(n_1, n_2 \ldots n_n) = p(n_1)p(n_2) \ldots p(n_n)$$

Example. What is the probability of throwing three sixes with three dice?

$$p = \tfrac{1}{6}.\tfrac{1}{6}.\tfrac{1}{6} = 0{\cdot}0046$$

Obviously this is rather unlikely and will occur on average approximately every 200 throws.

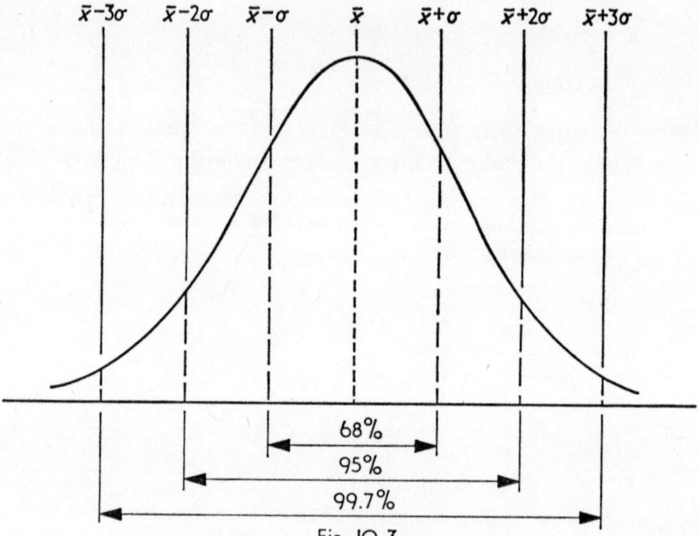

Fig. 10.3
The Normal Distribution

The normal curve allows the probability of any particular result occurring in a set of results to be calculated. Since the curve extends in theory from $-\infty$ to $+\infty$ the area under the curve may be equated with a probability, p, of unity. In other words,

$$\int_{-\infty}^{+\infty} f_{(x)}\,\mathrm{d}x = 1$$

Having made this normalisation we can now integrate the curve between any two values of x and find directly the probability of a result occurring in this region. If we wanted to know the probability of a result falling within one standard deviation, σ, from the mean, $\bar{x}$, this would be given by

$$p' = \frac{1}{\sigma\sqrt{(2\pi)}}\int_{\bar{x}-\sigma}^{\bar{x}+\sigma} \exp\left(-\tfrac{1}{2}\left(\frac{x-\bar{x}}{\sigma}\right)^2\right)\mathrm{d}x$$

$$\simeq 0{\cdot}68$$

Fortunately, due to its great value in statistical analysis, many tables of values for this integral are available so that the actual integration need not be carried out. Fig. 10.3, showing the normal distribution, indicates that approximately 68% of results lie within 1σ, 95% within 2σ and 99·7% within 3σ.

A simple table giving the probability, p, of finding a value outwith n standard deviations from the mean x is given in Table 10.1. (A more comprehensive table is given on page 140.)

Table 10.1.

n (σ)	p	n (σ)	p
0·0	1·0	2·0	0·046
0·2	0·841	2·2	0·028
0·4	0·689	2·4	0·016
0·6	0·549	2·6	0·0093
0·8	0·424	2·8	0·0051
1·0	0·317	3·0	0·0027
1·2	0·230	3·5	0·00047
1·4	0·162	4·0	0·000063
1·6	0·110	4·5	0·0000068
1·8	0·072	5·0	0·00000048

Approximately 10% of the results will be at $>1{\cdot}65\sigma$ from the mean; 5% at $>1{\cdot}96\sigma$; 1% at $>2{\cdot}58\sigma$ and 0·1% at $>3{\cdot}3\sigma$.

10.5 Confidence limits

If it were possible to make an infinite number of measurements then the arithmetic mean of these results would give the correct value. However this is not possible and situations often arise where only a few measurements are available. The arithmetic mean of a small set of results will rarely be the correct value and it is necessary therefore when quoting such a mean to indicate in some way the confidence one has (in a statistical sense) in the result obtained. Various statistical tests are available to deal with specific situations and three of these are discussed below.

We must at some stage decide what values of probability p we will accept as 'likely' or 'unlikely' or 'very unlikely'. This will of course be a fairly arbitrary decision but it is generally accepted that if the probability of an event occurring is less than 0·05 then its occurrence is *statistically significant*, i.e. the event is, by definition, 'unlikely' to have arisen by chance. Table 10.2 outlines the generally accepted interpretations of probability levels.

Table 10.2

Probability	*Interpretation*
$p \geqq 0{\cdot}1$	Not significant
$0{\cdot}1 > p \geqq 0{\cdot}05$	Possibly significant but doubtful
$0{\cdot}05 > p \geqq 0{\cdot}01$	Significant
$0{\cdot}01 > p \geqq 0{\cdot}001$	Highly significant
$p < 0{\cdot}001$	Very highly significant

The actual interpretation will always depend on the situation. It should be emphasised that one is always dealing with probabilities and that statistical analysis does not provide absolute proof. Results may be quoted in terms of *confidence limits*. For example, the 95% confidence limits give the range of values in which a result has an occurrance probability of 0·95.

10.6 Significance tests

A large number of specific statistical tests to deal with different situations have been devised and three of these are illustrated below.

(a) *ZM test*

This test compares a random sample of one or more measurements with a large parent group of which the mean, $\bar{X}$, and standard deviation, σ, are known

$$Z = \frac{\sqrt{n}|\bar{X} - \bar{x}|}{\sigma_{parent}}$$

where n = number of measurements in the sample

and $\bar{x}$ = arithmetic mean of the sample.

The probability of the sample belonging to the parent group may be found by substituting Z for n in Table 10.1.

Example. The results of a Cu^{2+} determination by a large student class had a mean of 18·9% Cu^{2+} and a standard deviation of 0·9%. Five subsequent analyses of supposedly the same solution gave a mean value of 20·0% Cu^{2+}.

Is there any statistical evidence that this is not the original solution given to the class?

Mean of sample, $\bar{x} = 20{\cdot}0$

Mean of parent group, $\bar{X} = 18{\cdot}9$

Standard deviation of parent group, $\sigma = 0{\cdot}9$

Number of analyses in sample, $n = 5$

$$Z = \frac{\sqrt{5}|18{\cdot}9 - 20{\cdot}0|}{0{\cdot}9}$$

$$= 2{\cdot}73$$

The probability of this result arising by chance is (see Table 10.1) approximately 0·007 i.e. less than 1 in 100. This is statistically highly significant and it is certainly worth having a look around the laboratory for another bottle.

(b) *Student's t test*

This test is a modified form of the *ZM* test and can be applied when only the mean and not the standard deviation of the large parent group is known. The random sample should contain at least 3 measurements.

$$t = \frac{\sqrt{n}|\bar{X} - \bar{x}|}{\sigma_{sample}}$$

The significance of the result may be found from a table relating t, n and probability (see page 141) An abbreviated form is given in Table 10.3.

Table 10.3

Degrees of freedom	t $p = 0{\cdot}10$	$p = 0{\cdot}05$	$p = 0{\cdot}01$
1	6·31	12·71	63·66
2	2·92	4·30	9·93
3	2·35	3·18	5·84
4	2·13	2·78	4·60
5	2·02	2·57	4·03
6	1·94	2·45	3·71
7	1·90	2·37	3·50
8	1·86	2·31	3·36

Example. The mean breaking tension of a particular type of nylon fibre derived from a large number of measurements was 5·0 kg. An unidentified fibre was then examined and it broke in five separate measurements at tensions of 4·9, 5·0, 5·6, 5·6 and 5·4 kg. Is there statistical evidence that this is not the original sample?

$$\text{Mean of sample } \bar{x} = 5{\cdot}3$$

$$\text{Standard deviation of sample} = 0{\cdot}33$$

$$t = \frac{\sqrt{5}|5{\cdot}0-5{\cdot}3|}{0{\cdot}33}$$

$$= 2{\cdot}03$$

The number of degrees of freedom is $n-1$ i.e. 4. From the table $t = 2{\cdot}03$ is above the 0·10 level and the conclusion is that there is no statistical difference between the two samples.

Example. A set of titration figures obtained by a student are 1·98, 1·97, 2·01, 1·96 and 1·98 ml. What are the 95% and 99% confidence limits of these results?

These limits are given by the formula

$$\bar{x} \pm \frac{t\sigma}{\sqrt{n}}$$

where $\bar{x}$ = mean result

n = number of results

σ = standard deviation of the results

and t = constant for $n-1$ degrees of freedom at the required confidence level, i.e. 0·05 and 0·01 in this case. Thus for the 95% confidence limits $t = 2{\cdot}78$ and for the 99% confidence limit $t = 4{\cdot}60$

$$\bar{x} = 1{\cdot}98$$

$$\sigma = 0{\cdot}019$$

$$n = 5.$$

At the 95% level the limits are

$$1{\cdot}98 \pm 2{\cdot}78(0{\cdot}019)/\sqrt{5} = 1.98 \pm 0{\cdot}024 \text{ ml}$$

At the 99% level the limits are

$$1{\cdot}98 \pm 4{\cdot}60(0{\cdot}019)/\sqrt{5} = 1{\cdot}98 \pm 0{\cdot}039 \text{ ml}$$

(c) χ^2 *test* (*chi-squared test*)

This test may be used to determine whether the frequency of occurrence of any particular event is significantly different from the expected frequency of occurrence. The value of χ^2 for a particular situation is given by

$$\chi^2 = \frac{\Sigma(f_{obs} - f_{exp})^2}{f_{exp}}$$

where f_{obs} = observed frequencies

and f_{exp} = expected frequencies

An abbreviated form of a typical χ^2 table is given in Table 10.4 (see also page 142).

Table 10.4

Degrees of freedom	χ^2 $p = 0{\cdot}1$	$p = 0{\cdot}05$	$p = 0{\cdot}01$
2	4·61	5·99	9·21
3	6·25	7·82	11·34
4	7·78	9·49	13·28
5	9·24	11·07	15·09
6	10·65	12·59	16·81
7	12·02	14·07	18·48
8	13·36	15·51	20·09
9	14·68	16·92	21·67
10	15·99	18·31	23·21

Example. The number of breakages was recorded in a laboratory over a prolonged period on a daily basis. The results were 61, 64, 57, 73 and 80 on Mondays, Tuesdays, Wednesdays, Thursdays and Fridays respectively. Do these results suggest that people are more accident-prone on specific days of the week?

Expected frequency, $f_{exp} = \Sigma f_{obs}/5 = 335/5 = 67$.

	f_{obs}	$(f_{obs} - f_{exp})$	$(f_{obs} - f_{exp})^2$
Monday	61	− 6	36
Tuesdays	64	− 3	9
Wednesday	57	−10	100
Thursday	73	+ 6	36
Friday	80	+13	169

$$\chi^2 = 350/67 = 5{\cdot}22$$

The example above has $(n-1)$ degrees of freedom (i.e. 4). $\chi^2 = 5{\cdot}22$ is well outside the $p = 0{\cdot}1$ level and therefore there is no statistical evidence to suggest that the number of breakages is associated with the day of the week. It is worth noting that without this analysis one might have concluded that accidents were more likely at the end of the week.

10.7 Regression analysis—least squares method of finding a best straight line

Previously we have considered only those situations involving one variable. However in many cases two variables have to be considered. For example, the results plotted in Fig. 10.4 show a certain amount of scatter and a common problem is to find the best straight line through these points. The method to be described is an example of regression analysis.

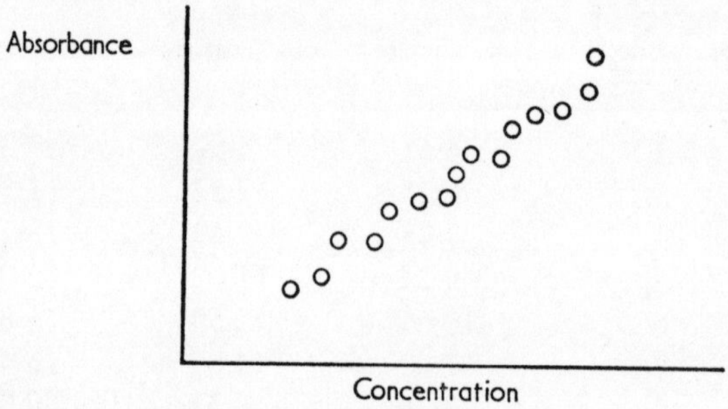

Fig. 10.4

This may be done by finding the line $y = mx+c$ for which the sum of the squares of the deviations from the line is a minimum. The deviation between a given point on the best straight line at $x = x_i$ i.e. $y_i' = a+bx_i$ and the experimental point at $x = x_i$ namely, y_i, is $y_i'-y_i$. We therefore require that

$$\sum_{i=1}^{n} (y_i'-y_i)^2 \text{ is a minimum.}$$

By taking the partial derivatives of this sum with respect to a and b and solving the resulting simultaneous equations we obtain,

$$m = \frac{n(\Sigma xy)-\Sigma x(\Sigma y)}{n(\Sigma x^2)-(\Sigma x)^2}, \quad \text{and}$$

$$c = \bar{y}+m\bar{x}$$

where all the sums are taken between 1 and n, and $\bar{y}$ and $\bar{x}$ are the arithmetic means of n results.

Example. Find the best straight line through the points (1, 0·8), (2, 1·2), (3, 2·8), (4, 3·0) and (5, 4·7)

x	y	xy	x^2
1	0·8	0·8	1
2	1·2	2·4	4
3	2·8	8·4	9
4	3·0	12·0	16
5	4·7	23·5	25

$$\Sigma x = 15{\cdot}0 \qquad \bar{x} = \Sigma x/5 = 3{\cdot}0$$
$$\Sigma y = 12{\cdot}5 \qquad \bar{y} = \Sigma y/5 = 2{\cdot}5$$
$$\Sigma xy = 47{\cdot}1$$
$$\Sigma x^2 = 55$$

$$m = \frac{5(47{\cdot}1)-15(12{\cdot}5)}{5(55)-(15)^2}$$

$$= 0{\cdot}96$$

$$c = 2{\cdot}5-0{\cdot}96(3{\cdot}0) = -0{\cdot}38$$

The best straight line is $y = -0{\cdot}38+0{\cdot}96x$.

10.8 Standard error and confidence limits of the regression coefficient, *m*

The *standard error* of the coefficient *m*, *S.E.*(*m*), may be found from the expression,

$$S.E.(m) = \sigma_0 \Big/ \sum_{i=1}^{n} (x_i-\bar{x})^2$$

$$\text{where } \sigma_0 = \frac{1}{n-2}\sum_{i=1}^{n} (y_i-c-mx_i)^2$$

Using this value of the standard error the confidence limits on the correct value of m may be found. The $100(1-\alpha)\%$ confidence limits are

$$m \pm t_{\frac{1}{2}\alpha}(S.E.(m))$$

where the value of t is obtained from a standard 't' table (see Section 10.6(b)) and the number of degrees of freedom is $n-2$ where n is the number of points.

10.9 Standard error and confidence limits of an estimated y value

The standard error of any value, y_0, predicted for a given value of x_0, $S.E.(y_0)$, is given by,

$$S.E.(y_0) = \sigma_0 . \left\{ 1/n + \frac{(x_0 - \bar{x})^2}{\sum_{i=1}^{n} (x_i - \bar{x})^2} \right\}^{\frac{1}{2}}$$

and the $100(1-\alpha)\%$ confidence limits on the estimate by

$$y_0 \pm t_{\frac{1}{2}\alpha}(S.E.(y_0))$$

using 't' tables and $n-2$ degrees of freedom.

10.10 Curvilinear regression

The procedures described in Sections 10.7, 10.8 and 10.9 apply to straight lines. However the general method may be extended to deal with *curves* where the relevant equations contain second or higher degree values of x, e.g.

$$y = lx^2 + mx + c$$

The sum of squares of the errors are now be

$$\sum_{i=1}^{n} (y_i - lx_i^2 - mx_i - c)^2.$$

Best values for l, m and c may be obtained in this quadratic case by solving the three simultaneous equations,

$$\Sigma y_i - cn - m\Sigma x_i - l\Sigma x_i^2 = 0$$

$$\Sigma x_i y_i - c\Sigma x_i - m\Sigma x_i^2 - l\Sigma x_i^3 = 0$$

$$\Sigma x_i^2 y_i - c\Sigma x_i^2 - m\Sigma x_i^3 - l\Sigma x_i^4 = 0$$

Analogous methods are used for higher polynomials.

10.11 Asymmetric distribution—mean, median and mode

In dealing with symmetric distributions we find that the arithmetic mean gives a very good indication of the 'centre' of the distribution and, if the number of results is sufficiently high, approximates to the correct result. However many distributions are not symmetrical and the arithmetic mean does not give a good indication of the 'centre'. Two other parameters, the *median* and the *mode*, are often used in these circumstances.

The median may be obtained by arranging the observations in ascending order of magnitude in which case the median is the observation with an equal number of observations above and below it if there is an odd number of observations, or the average of the two central values if there is an even number of observations.

The mode is simply the value with the maximum frequency. Distributions with a single mode are *unimodal* and they approximately obey the relation,

$$\text{Mean} - \text{mode} = 3(\text{mean} - \text{median}).$$

The degree of asymmetry or *skewness* of a distribution is sometimes represented by Pearson's measure of skewness which is given by

$$\text{Skewness} = \frac{(\text{mean} - \text{mode})}{\text{standard deviation}}$$

Some examples of mean, mode and median are shown in Fig. 10.5 (a, b and c).

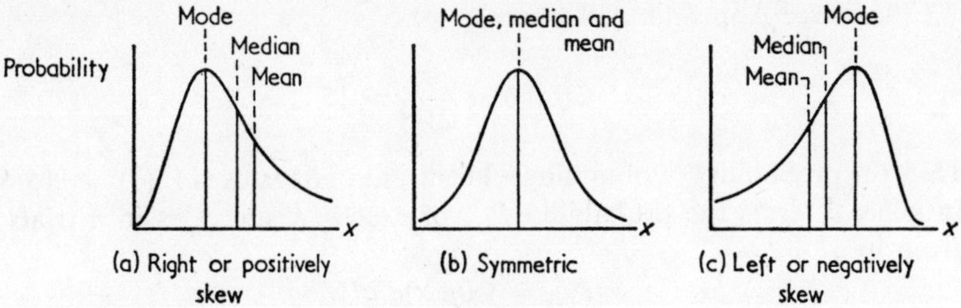

Fig. 10.5

It will be obvious the arithmetic mean, mode and median are identical for a symmetrical distribution (Fig. 10.5(b)).

10.12 Geometric, logarithmic and harmonic means

Mean values other than the arithmetic mean are sometimes used and three of these are briefly described below

(1) *Geometric mean*

$$\bar{x}_G = \sqrt[n]{(x_1 . x_2 . x_3 \ldots x_n)}$$

i.e. the nth root of the product of n values.

(2) *Logarithmic mean*

$$\bar{x}_L = \frac{1}{n} \sum_{i=1}^{n} \log x_i = \log \bar{x}_G$$

i.e. the logarithm of the geometric mean.

(3) *Harmonic mean*

$$\bar{x}_H = \frac{n}{\Sigma(1/x_i)}$$

i.e. the reciprocal of the mean of the reciprocals of the observations.

10.13 Binomial distribution

A situation that sometimes arises is that in which there are only two possible outcomes of an experiment. These we may call success (with probability p) and failure (with probability q) and since the result must be one or the other, $p+q = 1$. For example, when a coin is tossed the probability of obtaining heads (p) is equal to that of obtaining tails (q) as a result so that $p = q = \frac{1}{2}$. We may now consider the probability of obtaining say exactly 4 heads out of 6 tosses of the coin. The probability of obtaining a particular sequence such as *hhhtht* will be $(\frac{1}{2})^6$ (see Section 10.4) and the number of possible sequences containing 4 heads and 2 tails is $C(6, 4)$ (see Appendix A7) where

$$C(6, 4) = \frac{6!}{2!4!} = 15$$

Thus the probability of obtaining 4 heads out of 6 tosses is $15(\frac{1}{2})^6 = 15/64$. In general terms the probability $P_{(x)}$ of exactly x successes in n trials is given by

$$P_{(x)} = C(n, x)p^x q^{n-x}$$

and the probability of not more than x successes, which is the sum of the probabilities of 0, 1, 2, ... x successes, is given by

$$P'_{(x)} = C(n, 0)p^0 q^n + C(n, 1)p^1 q^{n-1} + \ldots C(n, x)p^x q^{n-x}$$

The terms above are terms of the binomial series obtained on expansion of $(p+q)^n$ and the function $P'_{(x)}$ is known as a *binomial distribution* function.

Example. A manufacturer sells thermometers in packs of 5 knowing that 5% of his stock is defective. What are the probabilities of a pack containing 0, 1, 2, 3, 4 or 5 standard thermometers?

p = probability of a thermometer being standard = 0·95

q = probability of a thermometer being defective = 0·05

n = number of trials = 5

Probability of pack containing x standard thermometers is given by

$$P_{(x)} = C(5, x)0{\cdot}95^x 0{\cdot}05^{5-x}.$$

x	$C(5, x)$	$0{\cdot}95^x$	$0{\cdot}05^{5-x}$	$P_{(x)}$
0	1	1	$3{\cdot}13\times10^{-7}$	3×10^{-7}
1	5	0·95	$6{\cdot}25\times10^{-6}$	3×10^{-5}
2	10	0·903	$1{\cdot}25\times10^{-4}$	0·001
3	10	0·857	$2{\cdot}5\times10^{-3}$	0·021
4	5	0·815	0·05	0·204
5	1	0·774	1	0·774

Thus on average 77% of the packs will contain no defective thermometers, 20% will contain one defective thermometer and 2% will contain two defective thermometers. The probability of obtaining a pack of 5 defective thermometers is extremely low: approximately 3 packs in every 10 million.

Application of the binomial distribution method to situations where n is large is obviously going to involve a considerable amount of computation and to alleviate this situation two other distributions are widely used. These are (a) the Poisson distribution which is used when p is very small (Section 10.14) and (b) the Normal distribution which we have already discussed (Section 10.3)—this may be used when both n and np are large.

10.14 Poisson distribution

This distribution may be derived from the binomial distribution

$$P_{(x)} = C(n, x)p^x q^{n-x}$$

by letting $n \to 0$, $p \to 0$ and the limit $np = \alpha$ where α is a finite number, and may be represented by

$$P_{(x)} = \frac{\alpha^x e^{-\alpha}}{x!}$$

where $P_{(x)}$ is the probability of an event occurring x times per unit time and α is the average number of events per unit time.

Example. The background count of a γ-ray detector over a period of 1000 minutes was 1360. If we now count a γ-emitting sample over periods of 30 seconds what are the probabilities that we will be including 0, 1, 2 and 3 counts caused by the background?

The average number of background counts per minute is 1·36 which will be 0·68 counts per 30 seconds. Therefore $\alpha = 0{\cdot}68$ and the probability of no background count during a 30 second interval is

$$P_{(0)} = \frac{0{\cdot}68^0 e^{-0{\cdot}68}}{1} = 0{\cdot}507.$$

Analogously

$$P_{(1)} = \frac{0{\cdot}68^1 e^{-0{\cdot}68}}{1} = 0{\cdot}68(0{\cdot}507) = 0{\cdot}345$$

$$P_{(2)} = \frac{0{\cdot}68^2 e^{-0{\cdot}68}}{2!} = 0{\cdot}117$$

and $$P_{(3)} = \frac{0{\cdot}68^3 e^{-0{\cdot}68}}{3!} = 0{\cdot}027$$

Therefore the probabilities of 0, 1, 2 and 3 counts are 0·567, 0·345, 0·117 and 0·027 respectively, during the 30 second intervals.

Both the binomial and Poisson distributions apply only to situations where there are only two possible outcomes. In the example above for instance there is either a count registered during a given time interval or there is not. The normal distribution, on the other hand, may be applied to continuous data i.e. where there are several possible results.

11 *Graphical Methods*

11.1 Graphs of experimental data

If one physical property of a system is a function of another, a set of experimental data may be formed by measuring the values of the former at a series of known values of the latter. These data may be plotted on a graph. It is customary to plot the *dependent variable* on the vertical or y-axis and the *independent variable* on the horizontal or x-axis. Thus Fig. 11.1 is a graph of the optical absorbance of a $K_2Cr_2O_7$ solution (the dependent variable) against the wavelength of the light absorbed (the independent variable).

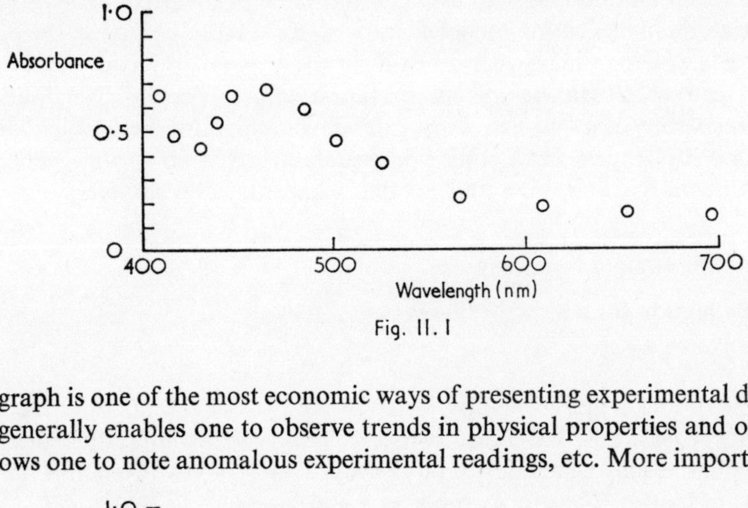

Fig. 11.1

A graph is one of the most economic ways of presenting experimental data. It generally enables one to observe trends in physical properties and often allows one to note anomalous experimental readings, etc. More important,

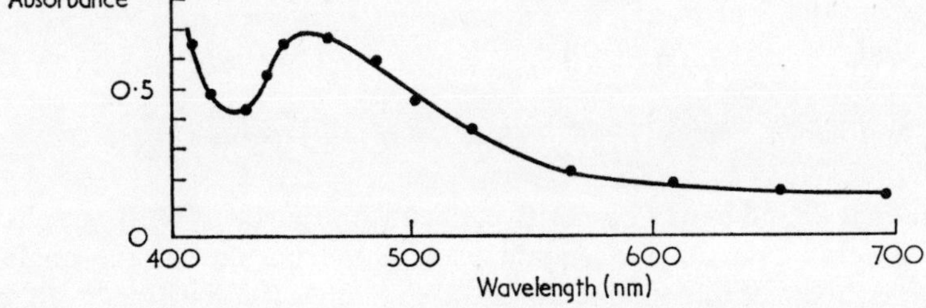

Fig. 11.2

however, a graph provides a simple and relatively precise method of obtaining intermediate values of the dependent variable by the process of *interpolation*. Thus by drawing a smooth curve through the points in Fig. 11.1 it becomes possible to estimate the value of the absorbance at a wavelength where no reading has been taken or, for example, the value of the wavelength corresponding to the maximum absorbance within the wavelength range of the experiment.

[Note that the smooth curve through the points represents the *function* which describes the variation of absorbance with wavelength, although the curve was drawn with no foreknowledge of the corresponding mathematical expression. In many cases chemists require to find such an equation: this can often be done by making a series of successive approximations to the function, usually using a computer, until a satisfactory curve-fit is achieved.]

It is sometimes necessary to estimate the value of the dependent variable outside the limits of the independent variable used in obtaining the data. For example one may wish to estimate the value of the equivalent conductance of a salt as its concentration approaches infinite dilution. *Extrapolation* of a curve involves extending it smoothly beyond its limits. Even if it is known from other evidence that there is no abrupt change in the direction of the curve, this is often a hazardous undertaking.

11.2 The straight line graph

The equation for a straight line is

$$y = mx+c$$

where m and c are constants; m is the gradient of the line and c is the intercept *on the y-axis*. These two constants which completely define the straight line may be determined if the coordinates of any two points on the line are known. Thus for two points 1 and 2 we have

$$y_1 = mx_1+c$$

and $$y_2 = mx_2+c$$

and hence $$m = \frac{y_2-y_1}{x_2-x_1} \quad \text{and} \quad c = \frac{x_2y_1-x_1y_2}{x_2-x_1}$$

Note that m and c may be positive, negative or zero. [N.B. In graphs it is not generally possible to equate m with the tangent of the angle the straight line makes with the x-axis. This is only true in isometric graphs where the axes are drawn to the same scale.]

Because of the simplicity of extrapolation and interpolation and the ease in finding the equation of the function, it is common practice to arrange experimental data so as to give straight line plots.

Example. The rate constant of a reaction k, may be given by the expression

$$k = A \exp(-E_a/RT)$$

where A, E_a and R are constants and T is the absolute temperature. Show that this relationship holds for the following data, determine A and E_a and find the value of the rate constant at 100°C.

Temperature (°C)	Rate constant ($\text{l mol}^{-1}\ \text{s}^{-1}$)
0	$1{\cdot}01 \times 10^{-5}$
20	$5{\cdot}62 \times 10^{-5}$
40	$2{\cdot}63 \times 10^{-4}$
60	$1{\cdot}00 \times 10^{-3}$
80	$2{\cdot}95 \times 10^{-3}$

Our first aim is to find suitable functions which would produce straight line plots if the theoretical expression holds. Taking logarithms of the latter we have

$$\ln k = \ln A + \ln(\exp(-E_a/RT))$$

i.e.

$$\ln k = \ln A + (-E_a/RT)$$

or

$$\log_{10} k = \log_{10} A - \frac{E_a}{2{\cdot}303R}\left(\frac{1}{T}\right)$$

If now $\log_{10} k$ is equated with y and $1/T$ with x it is seen that we have the equation

$$y = \log_{10} A - \left(\frac{E_a}{2{\cdot}303R}\right)x$$

i.e. an equation of the form $y = c + mx$ where the intercept on the y-axis is $\log_{10} A$ and the gradient has a value of $-E_a/2{\cdot}303R$.

k	$y = \log_{10} k$	T	$x = 1/T$
$1{\cdot}01 \times 10^{-5}$	$\bar{5}{\cdot}00 = -5{\cdot}00$	273	0·00366
$5{\cdot}62 \times 10^{-5}$	$\bar{5}{\cdot}75 = -4{\cdot}25$	293	0·00341
$2{\cdot}63 \times 10^{-4}$	$\bar{4}{\cdot}42 = -3{\cdot}58$	313	0·00319
$1{\cdot}00 \times 10^{-3}$	$\bar{3}{\cdot}00 = -3{\cdot}00$	333	0·00300
$2{\cdot}95 \times 10^{-3}$	$\bar{3}{\cdot}47 = -2{\cdot}53$	353	0·00286

[N.B. It is not possible to directly plot a number such as $\bar{5}{\cdot}75$. We recall that the meaning of $\bar{5}{\cdot}75$ is $(-5{\cdot}00+0{\cdot}75)$, i.e. $-4{\cdot}25$.]

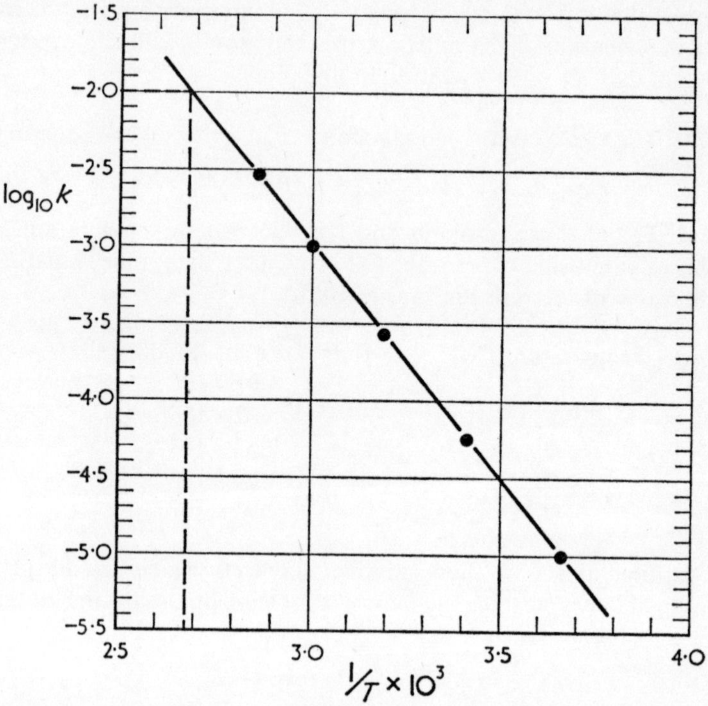

Fig. 11.3

A straight line graph is produced (Fig. 11.3) and we therefore conclude that the reaction kinetics conform to the given theoretical expression. The value of the slope is given by

$$m = \frac{y_2 - y_1}{x_2 - x_1} = \frac{\log_{10} k_2 - \log_{10} k_1}{(1/T_2) - (1/T_1)}$$

$$= \frac{4{\cdot}22 - (-2{\cdot}68)}{0{\cdot}00340 - 0{\cdot}00290} = \frac{-1{\cdot}54}{0{\cdot}00050} = -3080{\cdot}0.$$

Hence $-E_a/2{\cdot}303R = -3080$ and $E_a = 59$ kJ mol^{-1} if we take $R = 8{\cdot}314$ JK^{-1} mol^{-1}

Also
$$c = \frac{x_2 y_1 - x_1 y_2}{x_2 - x_1}$$

$$= \frac{0{\cdot}00340(-2{\cdot}68) - 0{\cdot}00290(-4{\cdot}22)}{0{\cdot}00340 - 0{\cdot}00290}$$

$$= 6{\cdot}2$$

therefore

$$\log_{10} A = 6{\cdot}2 \quad \text{and} \quad A = 1{\cdot}59 \times 10^6.$$

Finally by graphical extrapolation (or by calculation from the expression $\log_{10} k = -3080(1/T)+6{\cdot}2$) we have

$$\log_{10} k_{373} = -2{\cdot}0$$

and thus the rate constant at 100°C has a value of 0·01 l mol^{-1} s^{-1}.

Example. The Jones-Dole equation relating the relative viscosity, η, of an electrolyte solution to the concentration, c, of the electrolyte has the form

$$\eta = 1+A\sqrt{c}+Bc$$

where A and B are constants. Suggest a function of η and c which might be plotted to verify this relationship for a particular experiment.

$$\eta = 1+A\sqrt{c}+Bc$$

therefore $$\eta-1 = A\sqrt{c}+Bc$$

therefore $$\left(\frac{\eta-1}{\sqrt{c}}\right) = A+B\sqrt{c}$$

Thus a plot of $(\eta-1)/\sqrt{c}$ against $\sqrt{c}$ should be drawn.

11.3 Triangular graphs

On application of the phase rule it is found that a ternary (three component) system, under conditions of constant temperature and pressure, has two degrees of freedom. If the system consists of a single phase, these variables may be taken as two of the three mole fractions x_1 and x_2 of components 1 and 2 respectively. The third mole fraction is then fixed since $x_1+x_2+x_3 = 1$. The composition of such a system is conveniently represented by the position of a point in an equilateral triangle where the mole fraction of each component is plotted on a normal to one side of the triangle.

Thus if we consider a homogeneous system containing components A, B and C having mole fractions 0·3, 0·5 and 0·2 respectively, the mole fraction of A is represented by any point on the heavy line shown in Fig. 11.4: i.e. the point representing the composition of the system must lie somewhere on this line.

Similarly mixtures with $x_B = 0{\cdot}5$ are represented by the heavy line in Fig. 11.5

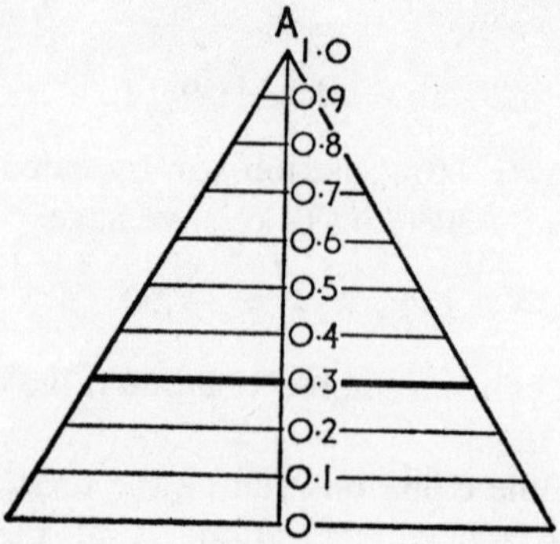

Fig. 11.4

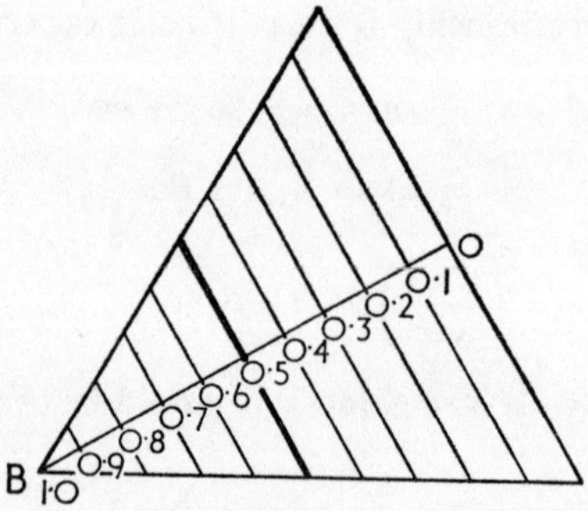

Fig. 11.5

The composition of the system must then be fixed by the point p where the two heavy lines cross:

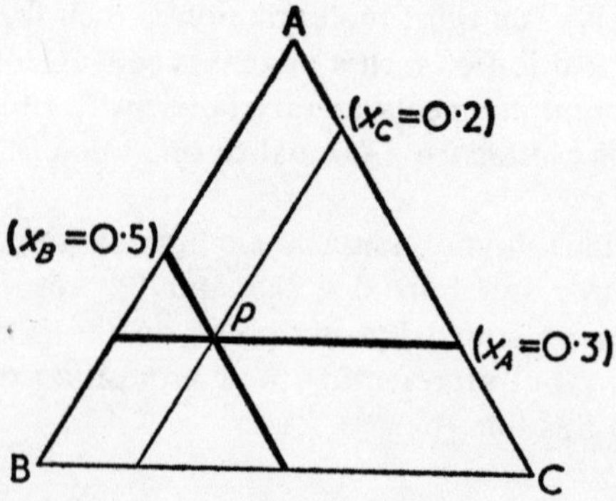

Fig. 11.6

Now the *sum of the lengths of the three perpendiculars drawn from any point within an equilateral triangle is always equal to the length of one side of the triangle.* Hence as may be shown by the geometry of equivalent triangles this point must also be intersected by the $x_C = 0{\cdot}2$ line. Thus the point *p* defines the composition of this particular system. Triangular graphs have a further useful property. In Fig. 11.7 points on the line *AD* represent mixtures where the mole fractions of *B* and *C* are always in the fixed ratio

$$\frac{x_B}{x_C} = \frac{DC}{BD}$$

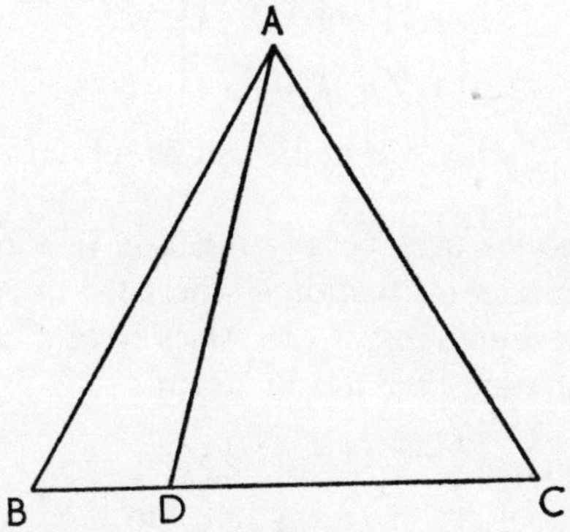

Fig. 11.7

[In some practical situations the three components are not mutually soluble over the entire composition range. Under these circumstances a two-phase region appears: within this area the above arguments do not hold.]

11.4 Nomograms

A simple *nomogram* or *alignment chart* consists of three *scales* corresponding to three interrelated variables *p*, *q* and *r*. If any two of these variables are known, the third can be read from the nomogram by drawing a straight line through the two known points and finding where it intersects the third scale. Such diagrams are extremely useful where calculations are to be performed repeatedly over a period of time and the method has found wide application.

Consider three points on the *xy* plane, $A(x_1, y_1)$, $B(x_2, y_2)$ and $C(x_3, y_3)$. The gradient of the line joining *AB* is $(y_2-y_1)/(x_2-x_1)$ and of that joining *AC* is $(y_3-y_1)/(x_3-x_1)$. Now if

$$\frac{y_2-y_1}{x_2-x_1} = \frac{y_3-y_1}{x_3-x_1}$$

A, *B* and *C* must lie on the same straight line.

This condition of collinearity may be expressed in determinant form as

$$\begin{vmatrix} x_1 & y_1 & 1 \\ x_2 & y_2 & 1 \\ x_3 & y_3 & 1 \end{vmatrix} = 0.$$

If now the coordinates of A are a function of p, those of B a function of q and those of C a function of r and these can be written in the form

$$\begin{vmatrix} x_1(p) & y_1(p) & 1 \\ x_2(q) & y_2(q) & 1 \\ x_3(r) & y_3(r) & 1 \end{vmatrix} = 0$$

then the expression relating p, q and r can be set out as a nomogram.

Each of the three scales may be any function in the xy plane, but the simplest and most common situation is where the three scales are straight and parallel, i.e. the equations of the scales have a constant value of x. The determinant equation then has the form

$$\begin{vmatrix} -c & y_1(p) & 1 \\ 0 & y_2(q) & 1 \\ c & y_3(r) & 1 \end{vmatrix} = 0$$

where c is a constant and represents the horizontal spacing between the scales. Expanding this determinant we have

$$y_1(p) - 2y_2(q) + y_3(r) = 0$$

As an example of such a situation we may consider the equation relating the pH of a buffer solution containing $[HA]$ mol l^{-1} of a weak acid and $[A^-]$ mol l^{-1} of its salt. The pH of the buffer solution is related to the pK of the weak acid by the expression

$$\text{pH} = \text{p}K_a + \log_{10} [A^-]/[HA]$$

This may be written in the form

$$\text{pH} - \text{p}K_a + \log \frac{[HA]}{[A^-]} = 0$$

and we can put

$$y_1 = \text{pH}$$

$$y_2 = \text{p}K_a/2$$

$$y_3 = \log [HA]/[A^-]$$

and thus satisfy the expression

$$y_1 - 2y_2 + y_3 = 0$$

Setting out the three scales on the plane we have the situation shown in Fig. 11.8.

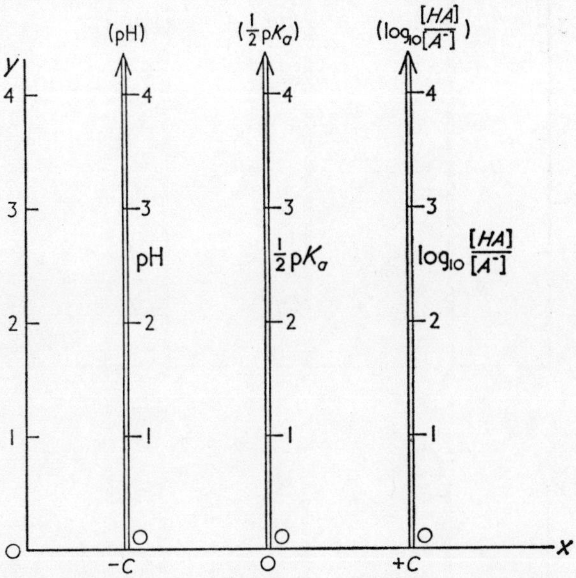

Fig. 11.8

This diagram is not particularly useful since it only deals with solutions where $[HA] > [A^-]$. In general it proves most convenient to adjust the zero of the scales in order to bring useful ranges of the variables onto the chart. Thus in the example considered above another valid set of scales is formed by setting

$$y_1 = -1{\cdot}5 + \text{pH}$$

$$y_2 = 0{\cdot}5 + \text{p}K_a/2$$

and $$y_3 = 2{\cdot}5 + \log [HA]/[A^-]$$

The result is shown in Fig. 11.9.

This chart might be given a more suitable notation as shown in Fig. 11.10.

The line drawn on the nomogram shows that the buffer formed from a weak acid with p$K_a = 3{\cdot}0$ and having $[HA]/[A^-] = 0{\cdot}1$, has a pH of 4·0.

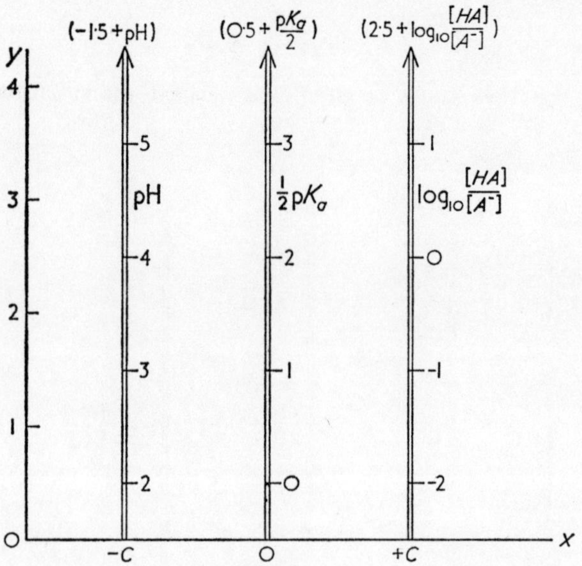

Fig. 11·9

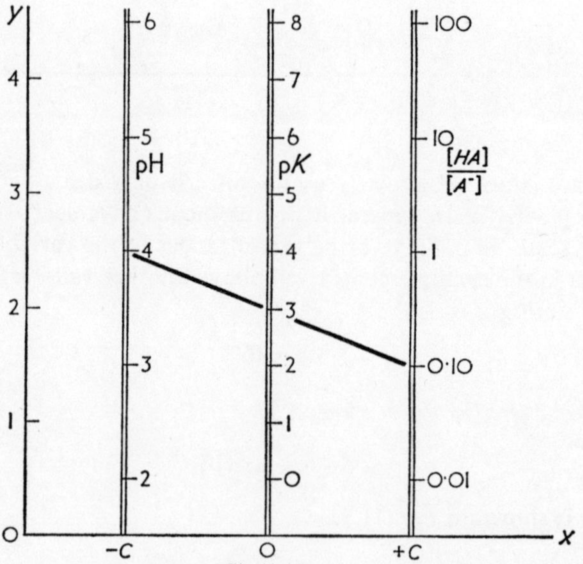

Fig. 11.10

In more involved situations the equations of the scales are functions of both x and y.

Appendices

A *Basic Equations*

A.1 Indices

1. $a^m = \underbrace{a.a.a. \dots a}_{m \text{ factors}}$
2. $a^m.a^n = a^{m+n}$
3. $a^m/a^n = a^{m-n}$
4. $(a^m)^n = a^{mn}$

Also $a^0 = 1$; $\quad a^{-m} = 1/a^m$; $\quad a^{1/m} = \sqrt[m]{a}$

Since chemists often require to use very large or very small numbers, it is convenient to express them in *semiexponential* form:

e.g.	*Number*	*Semiexponential number*
	96 490	$9{\cdot}649 \times 10^4$
	0·05916	$5{\cdot}916 \times 10^{-2}$

Care must be taken in manipulating such numbers:

e.g. (a) $(1{\cdot}2 \times 10^{-4}) \times (1{\cdot}5 \times 10^3) = (1{\cdot}2 \times 1{\cdot}5) \times 10^{(-4+3)}$
$$= 1{\cdot}8 \times 10^{-1}$$

(b) $(1{\cdot}3060 \times 10^{-3}) + (1{\cdot}306 \times 10^{-4})$
$$= 1{\cdot}3060 \times 10^{-3} + 0{\cdot}1306 \times 10^{-3} = 1{\cdot}4366 \times 10^{-3}$$

A.2 Logarithms

1. If $x = a^m$, then $\log_a x = m$
2. $\log_a (xy) = \log_a x + \log_a y$
3. $\log_a (x/y) = \log_a x - \log_a y$
4. $\log_a x^m = m \log_a x$

Also $\log_a a = 1$; $\log_a 1 = 0$

N.B. Equations involving powers can always be simplified by taking logarithms.

Example. Find the value of $5^{\frac{2}{3}}$.

$$\log (5^{\frac{2}{3}}) = \tfrac{1}{3} \log 5 = \tfrac{1}{3}(0{\cdot}699) = 0{\cdot}233$$

therefore $$5^{\frac{2}{3}} = 1{\cdot}71$$

5. $\log_a n = \log_a b . \log_b n$

therefore $$\log_e n (= \ln n) = \log_e 10 . \log_{10} n$$
$$= 2{\cdot}30258 \ldots \log_{10} n$$

Also $\log_a b . \log_b a = 1$

6. Note that the semiexponential form of a number is ideal for taking logarithms:

e.g. if $$x_1 = 9{\cdot}649 \times 10^4$$
$$\log_{10} x_1 = \log_{10} 9{\cdot}649 + \log_{10} 10^4$$
$$= 0{\cdot}9845 + 4$$
$$= 4{\cdot}9845$$

and if $$x_2 = 5{\cdot}916 \times 10^{-2}$$
$$\log_{10} x_2 = \log 5{\cdot}916 + \log_{10} 10^{-2}$$
$$= 0{\cdot}7720 + (-2)$$
$$= \bar{2}{\cdot}7720 \text{ (or } -1{\cdot}2280)$$

The index of the ten becomes the 'characteristic' of the logarithm.

Example. The pH of a solution may be taken approximately as $-\log_{10} [H^+]$ where $[H^+]$ is the molar hydrogen ion concentration. What is (a) the pH of a solution where $[H^+] = 0{\cdot}00012$, and (b) the hydrogen ion concentration in a solution which has a pH of 8·70?

(a) $[H^+] = 0{\cdot}00012 = 1{\cdot}2 \times 10^{-4}$

$$\log_{10} [H^+] = \bar{4}{\cdot}08 = -4{\cdot}00 + 0{\cdot}08$$
$$= -3{\cdot}92$$

therefore $$\text{pH} = 3{\cdot}92$$

(b) pH = 8·70

$$\log_{10}[H^+] = -8{\cdot}70$$
$$= -9{\cdot}00+0{\cdot}30$$
$$= \bar{9}{\cdot}30$$

therefore $[H^+] = 2{\cdot}0\times 10^{-9}$

A.3 Exponentials

1. $\exp x = e^x$, where $e = 2{\cdot}71828 \ldots$

2. $\log_e(\exp x) = \ln(\exp x) = x$

A.4 Sums, etc.

1. $\sum_{r=1}^{n} a_r = a_1+a_2+a_3+\ldots+a_n$

Example. N particles are distributed such that N_1 are placed in an energy level E_1, N_2 in energy level E_2 ... and N_i in level E_i. What is the mean energy of a particle?

$$N = \sum_{r=1}^{i} N_r$$

Total energy $= \sum_{r=1}^{i} N_r E_r$

and the mean energy $= \sum_{r=1}^{i} N_r E_r \Big/ \sum_{r=1}^{i} N_r$

2. $\prod_{r=1}^{n} a_r = a_1.a_2.a_3 \ldots a_n$

3. $a! = a(a-1)(a-2) \ldots 3.2.1$ = 'factorial a'

By definition $0! = 1$.

A.5 Trigonometric Functions

The angle x may be expressed in either degrees or radians.

$180° = \pi$ radians.

From Fig. A.1

1. $\sin x = \dfrac{1}{\operatorname{cosec} x} = \dfrac{CB}{AC}$

2. $\cos x = \dfrac{1}{\sec x} = \dfrac{AB}{AC}$

3. $\tan x = \dfrac{1}{\cot x} = \dfrac{CB}{AB}$

4. $\sin x = \cos (90-x)^\circ = \sin (180-x)^\circ$

5. $\cos x = \sin (90-x)^\circ = -\cos (180-x)^\circ$

6. $\tan x = \cot (90-x)^\circ = -\tan (180-x)^\circ$

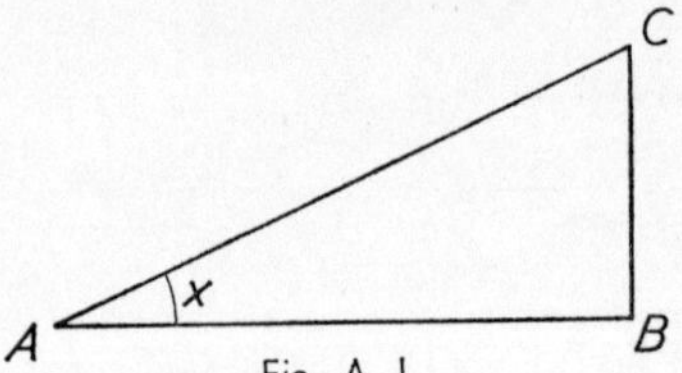

Fig. A.1

When angles are expressed in radians;

$$\sin x = \cos\left(\frac{\pi}{2}-x\right) = \sin (\pi - x), \text{ etc.}$$

7. $\cos (x+y) = \cos x \cos y - \sin x \sin y$

$\cos (x-y) = \cos x \cos y + \sin x \sin y$

These two relationships may be written in abbreviated form as $\cos (x\pm y) = \cos x \cos y \mp \sin x \sin y$

8. $\sin (x\pm y) = \sin x \cos y \pm \cos x \sin y$

9. $\tan (x\pm y) = \dfrac{\tan x \pm \tan y}{1\mp \tan x \tan y}$

10. $\sin 2x = 2 \sin x \cos x$

11. $\cos 2x = \cos^2 x - \sin^2 x$

12. $\tan 2x = \dfrac{2 \tan x}{(1-\tan^2 x)}$

13. $\sin x/\cos x = \tan x$

14. $\sin^2 x + \cos^2 x = 1$

15. $\sec^2 x - \tan^2 x = 1$

16. $\operatorname{cosec}^2 x - \cot^2 x = 1$

17. If $x = \sin y$

    $y = \sin^{-1} x = \text{arc} \sin x$

A.6 Hyperbolic Functions

1. $\cosh x = \frac{1}{2}(\exp x + \exp(-x))$
2. $\sinh x = \frac{1}{2}(\exp x - \exp(-x))$
3. *Osborne's rule*: If a formula can be written relating general trigonometric functions, then a corresponding formula may be written for hyperbolic functions provided that the sign of every product or implied product of two *sines* is reversed.

   Example 1. $\cos^2 x + \sin^2 x = 1$

   therefore $\cosh^2 x - \sinh^2 x = 1$

   Example 2. $\cos^2 x - \sin^2 x = \cos 2x$

   therefore $\cosh^2 x + \sinh^2 x = \cosh 2x$

   Example 3. $\sin x/\cos x = \tan x$.

   therefore $\sinh x/\cosh x = \tanh x$.

A.7 Permutations and Combinations

1. A *permutation* is an *arrangement* obtained by taking some or all of a number of objects.
2. A *combination* is a *selection* obtained by taking some or all of a number of objects.

   (Thus the possible permutations of 2 letters from *a b c* are *ab*, *ac*, *ba*, *bc*, *ca* and *cb* whereas the possible combinations are *ab*, *ac* and *bc* because, in the latter case, the ordering is not important.)
3. The number of permutations of x objects from a group of n different objects, $P(n, x)$, (sometimes written as ${}_nP_x$) is given by

$$P(n, x) = \frac{n!}{(n-x)!}$$

4. $P(n, n)$, the number of ways of arranging n different items is $n!$
5. The number of ways of arranging n objects where there are p objects of one type, q of another, etc., is

$$\frac{n!}{p!q! \ldots}$$

6. The number of combinations of x objects from a group of n different objects, $C(n, x)$, (sometimes written ${}_nC_x$) is given by

$$C(n, x) = \frac{P(n, x)}{x!} = \frac{n!}{(n-x)!x!}$$

7. It follows that

$$C(n, x) = C(n, n-x).$$

8. *Stirling's formula.* To avoid computation of $n!$ for large values of n Stirling's approximation,

$$n! \simeq n^n e^{-n} \sqrt{(2\pi n)},$$

or more simply,

$$\ln n! \simeq n \ln n - n, \text{ may be used.}$$

See also Table on page 144.

A.8 Series expansions (see chapter 9)

1. $\sin x = x - x^3/3! + x^5/5! - x^7/7! + \ldots$
2. $\cos x = 1 - x^2/2! + x^4/4! - x^6/6! + \ldots$
3. $e^x = \exp x = 1 + x/1! + x^2/2! + x^3/3! + \ldots$
4. $e = \exp 1 = 1 + 1/1! + 1/2! + 1/3! + \ldots$

   $= 2{\cdot}71828 \ldots$

B *Derivatives*

(see chapter 2)

u, v and w are functions of x; a is a constant.

$y = a$	$\mathrm{d}y/\mathrm{d}x = 0$
$y = au$	$\mathrm{d}y/\mathrm{d}x = a\,\mathrm{d}u/\mathrm{d}x$
$y = x^n$	$\mathrm{d}y/\mathrm{d}x = nx^{n-1}$
$y = u+v-w$	$\mathrm{d}y/\mathrm{d}x = \mathrm{d}u/\mathrm{d}x+\mathrm{d}v/\mathrm{d}x-\mathrm{d}w/\mathrm{d}x$
$y = u.v$	$\mathrm{d}y/\mathrm{d}x = u\,\mathrm{d}v/\mathrm{d}x+v\,\mathrm{d}u/\mathrm{d}x$
$y = \dfrac{u}{v}$	$\mathrm{d}y/\mathrm{d}x = \dfrac{v\,\mathrm{d}u/\mathrm{d}x-u\,\mathrm{d}v/\mathrm{d}x}{v^2}$
$y = e^{ax}$	$\mathrm{d}y/\mathrm{d}x = a\,e^{ax}$
$y = a^x$	$\mathrm{d}y/\mathrm{d}x = a^x \ln a$
$y = \ln x$	$\mathrm{d}y/\mathrm{d}x = 1/x$
$y = \sin ax$	$\mathrm{d}y/\mathrm{d}x = a \cos ax$
$y = \cos ax$	$\mathrm{d}y/\mathrm{d}x = -a \sin ax$
$y = \tan ax$	$\mathrm{d}y/\mathrm{d}x = a \sec^2 ax$
$y = \cot ax$	$\mathrm{d}y/\mathrm{d}x = -a\,\mathrm{cosec}^2\,ax$
$y = \sec ax$	$\mathrm{d}y/\mathrm{d}x = a \sec ax \tan ax$
$y = \mathrm{cosec}\,ax$	$\mathrm{d}y/\mathrm{d}x = -a\,\mathrm{cosec}\,ax \cot ax$

Appendix C lists a more complete range of standard integrals: derivatives can be evaluated by reversing the process of integration, e.g.

$$\int \frac{1}{a^2+x^2}\,\mathrm{d}x = \frac{1}{a}\tan^{-1}\frac{x}{a} \quad \text{therefore} \quad \mathrm{d}/\mathrm{d}x\left(\tan^{-1}\frac{x}{a}\right) = \frac{a}{a^2+x^2}.$$

It is also useful to note that for continuous functions

$$\mathrm{d}x/\mathrm{d}y = \frac{1}{\mathrm{d}y/\mathrm{d}x} \quad \text{and} \quad \mathrm{d}y/\mathrm{d}x = (\mathrm{d}y/\mathrm{d}u)(\mathrm{d}u/\mathrm{d}x)$$

C *Standard Integrals*

(see chapter 3)

The constant of integration is assumed throughout

$\int ax^n \, dx = (ax^{n+1}/n+1)$ except when $n = -1$

$\int (ax+b)^n dx = [(ax+b)^{n+1}/a(n+1)]$ except when $n = -1$

$\int e^{ax} \, dx = (1/a) \, e^{ax}$

$\int \sin ax \, dx = -(1/a) \cos ax$

$\int \tan ax \, dx = -(1/a) \ln (\cos ax)$

$\int \sec ax \, dx = (1/a) \ln \left[\tan \left(\frac{\pi}{4} + \frac{ax}{2} \right) \right]$

$\int \cos^2 x \, dx = \frac{1}{4}(2x + \sin 2x)$

$\int \tan^2 x \, dx = \tan x - x$

$\int \sec^2 x \, dx = \tan x$

$\int \sinh x \, dx = \cosh x$

$\int \tanh x \, dx = \ln (\cosh x)$

$\int (1/\sinh x) \, dx = \ln [\tanh (x/2)]$

$\int \cosh^2 x \, dx = \frac{1}{4}(2x + \sinh 2x)$

$\int \tanh^2 x \, dx = x - \tanh x$

$\int \mathrm{sech}^2 x \, dx = \tanh x$

$\int ax^{-1} \, dx = (a/n) \, dx = (1/a) \ln ax$

$\int 1/(ax+b) \, dx = (1/a) \ln (ax+b)$

$\int a^x \, dx = a^x (1/\ln a)$

$\int \cos ax \, dx = (1/a) \sin ax$

$\int \cot ax \, dx = (1/a) \ln (\sin ax)$

$\int \mathrm{cosec} \, ax \, dx = (1/a) \ln [\tan (ax/2)]$

$\int \sin^2 x \, dx = \frac{1}{4}(2x - \sin 2x)$

$\int \cot^2 x \, dx = -\cot x - x$

$\int \mathrm{cosec}^2 x \, dx = -\cot x$

$\int \cosh x \, dx = \sinh x$

$\int \coth x \, dx = \ln (\sinh x)$

$\int (1/\cosh x) \, dx = 2 \tan^{-1} e^x$

$\int \sinh^2 x \, dx = -\frac{1}{4}(2x - \sinh 2x)$

$\int \coth^2 x \, dx = x - \coth x$

$\int \mathrm{cosech}^2 x \, dx = -\coth x$

$\int \frac{1}{a^2 - x^2} \, dx = \frac{1}{a} \tanh^{-1} \frac{x}{a}$ if $x^2 < a^2$

$$\int \frac{1}{a^2+x^2}\,dx = \frac{1}{a}\tan^{-1}\frac{x}{a}$$

$$\int \frac{1}{x^2-a^2}\,dx = \frac{-1}{a}\coth^{-1}\frac{x}{a} \quad \text{if } a^2 < x^2$$

$$\int \frac{1}{\sqrt{(a^2+x^2)}}\,dx = \sinh^{-1}\frac{x}{a}$$

$$\int \frac{1}{\sqrt{(x^2-a^2)}}\,dx = \cosh^{-1}\frac{x}{a} \quad \text{if } a^2 < x^2$$

$$\int \frac{1}{\sqrt{(a^2-x^2)}}\,dx = \sin^{-1}\frac{x}{a} \quad \text{if } x^2 < a^2$$

$$\int \ln x\,dx = x(\ln x - 1)$$

Values of some definite integrals

$$\int_0^\infty x^n e^{-ax}\,dx = \frac{n!}{a^{n+1}}, \qquad n = 0, 1, 2 \ldots$$

$$\int_0^\infty e^{-ax^2}\,dx = \frac{1}{2}\left(\frac{\pi}{a}\right)^{\frac{1}{2}}$$

$$\int_0^\infty x^{2n} e^{-ax^2}\,dx = \frac{1.3.5\ldots(2n-1)}{2^{n+1}a^n}\left(\frac{\pi}{a}\right)^{\frac{1}{2}}, \qquad n = 1, 2 \ldots$$

$$\int_0^\infty x^{2n+1} e^{-ax^2}\,dx = \frac{n!}{2a^{n+1}}, \qquad n = 0, 1, 2 \ldots$$

Special functions

*Gamma function**

$$\Gamma(n) = \int_0^\infty e^{-x} x^{n-1}\,dx$$

*Error function**

$$\operatorname{erf} x = \frac{1}{\sqrt{\pi}}\int_0^x e^{-z^2}\,dz$$

$$\operatorname{erfc} x = 1 - \operatorname{erf} x$$

*Beta function**

$$B(p, q) = \int_0^1 x^{p-1}(1-x)^{q-1}\,dx \quad \text{for } p > 0, q > 0$$

---

* Values of these functions are available in standard mathematical tables.

D *Laplace Transforms*

(see chapter 4)

n is a positive integer; a is a constant.

FUNCTION $y = f(x)$	LAPLACE TRANSFORM $L(y)$ or $Y = F(p)$	
1	$1/p$	for $p > 0$
x^n	$n!/p^{n+1}$	,,
(e.g. x^2	$2/p^3$)	
$\sin ax$	$\dfrac{a}{p^2+a^2}$	,,
$\cos ax$	$\dfrac{p}{p^2+a^2}$	,,
$\exp ax$	$\dfrac{1}{p-a}$	for $p > a$
$\sinh ax$	$\dfrac{a}{p^2-a^2}$	,,
$\cosh ax$	$\dfrac{p}{p^2-a^2}$	,,
$x^n \exp ax$	$\dfrac{n!}{(p-a)^{n+1}}$	,,
$(1-ax)\exp(-ax)$	$\dfrac{p}{(p+a)^2}$	for $(p+a) > 0$

The above represent some of the simpler transforms. Extensive lists of transforms are available in standard mathematical tables.

E *Solutions of Polynomial Equations*

E.1 Quadratic equations

The general quadratic equation

$$ax^2+bx+c = 0$$

has two *roots* or solutions given by the 'quadratic formulae'

$$x_1 = \frac{-b+\sqrt{(b^2-4ac)}}{2a}$$

and $$x_2 = \frac{-b-\sqrt{(b^2-4ac)}}{2a}$$

so that numerical solutions may be obtained by substituting values of the coefficients a, b and c in these expressions. Usually only one of the roots is physically meaningful.

Great care must be taken in using these expressions when the values of a is small compared with b and c. In this situation the formulae may lead to inaccurate results and it is better to neglect the quadratic term and solve the resulting linear equation

$$bx+c = 0$$

E.2 Polynomial equations of higher degree

While formulae for solving cubic equations can be found, it is generally most convenient to find solutions to cubic and higher degree polynomials by *numerical methods* which permit calculation of a root to any required accuracy. These methods usually require an initial guess at the solution, followed by a series of *successive approximations*. 'Iterative' procedures of this nature tend to involve considerable amounts of repetitive computation and are often most efficiently carried out by electronic computers.

(a) *Method of successive approximations*

If an equation can be written in the form

$$x = \mathrm{f}(x)$$

a solution may sometimes be obtained by the method below. An initial value of the root is guessed and substituted in the right hand side of the equation thus permitting a new value of x to be determined. This in turn is substituted into $f(x)$ so that a further value of x is found.

i.e.
$$x_{n+1} = f(x_n)$$

Under suitable circumstances the value of x *converges* to the solution.
Example. Find the root of the equation

$$2x^2+x-2 = 0$$

by the method of successive approximations.

The equation may be rearranged to give

$$x = f(x) = \sqrt{(1-x/2)}$$

Let
$$x_1 = 0{\cdot}600$$

then
$$x_2 = f(x_1) = \sqrt{(1-0{\cdot}300)} = 0{\cdot}837$$
$$x_3 = f(x_2) = \sqrt{(1-0{\cdot}418)} = 0{\cdot}763$$
$$x_4 = f(x_3) = \sqrt{(1-0{\cdot}381)} = 0{\cdot}787$$
$$x_5 = f(x_4) = \sqrt{(1-0{\cdot}393)} = 0{\cdot}779$$
$$x_6 = f(x_5) = \sqrt{(1-0{\cdot}390)} = 0{\cdot}781$$
$$x_7 = f(x_6) = \sqrt{(1-0{\cdot}391)} = 0{\cdot}781$$

A solution to $2x^2+x-2 = 0$ is therefore $x = 0{\cdot}781$ (correct to the third place) as may be verified by substitution.

However not all functions can be treated successfully in this way since successive x values may be found to *diverge* from any initial value chosen. The *criterion for convergence* is that the numerical value of the first derivative of $f(x)$ is less than unity at the value of x corresponding to the solution.

(In the above example

$$f(x) = (1-x/2)^{\frac{1}{2}}$$

therefore
$$f'(x) = \frac{1}{4(1-x/2)^{\frac{1}{2}}} = 0{\cdot}32 \quad \text{where } x = 0{\cdot}781).$$

Because the convergence criterion is so restrictive and also because convergence is often very slow, this simple method of successive approximations is not always useful.

(b) *Newton's Method*

In Fig. E.1 the heavy line represents the function $y = f(x)$ and x_1 represents a guessed value of the root of the equation

$$y = f(x) = 0$$

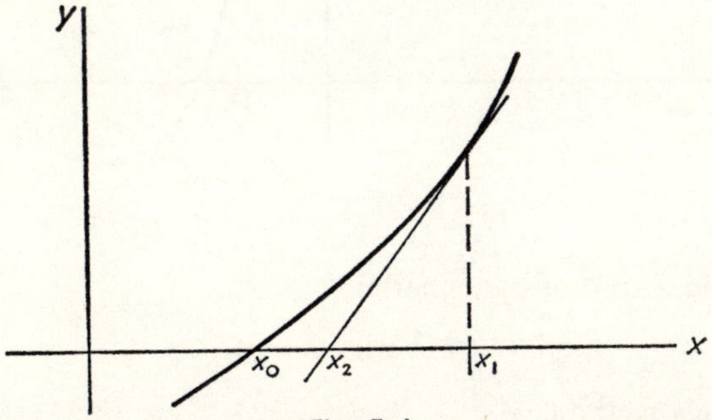

Fig. E.1

A tangent drawn to the curve at x_1 has a slope of $f'(x_1)$ and intersects the x-axis at x_2. It is seen that x_2 is closer to the true root x_0 than is x_1.

Further, by geometry,

$$f'(x_1) = \frac{f(x_1)-0}{x_1-x_2}$$

i.e.

$$x_2 = x_1 - \frac{f(x_1)}{f'(x_1)}$$

In general

$$x_{n+1} = x_n - \frac{f(x_n)}{f'(x_n)}$$

The criterion for convergence is that neither $f'(x)$ nor $f''(x)$ are zero at the value of x corresponding to the solution—i.e. $f'(x_0)$ and $f''(x_0)$ are non-zero. This condition imposes little restriction on the general use of the method. Care must be taken however in making the initial guess. Thus for example in Fig. E.2 choice of an initial value of x greater than x_1, would lead to an attempted convergence on x_2 rather than x_0. Similarly in Fig. E.3 an initial choice of x_1 would lead to convergence on the root $(x_0)_1$ whereas a choice of x_2 would lead to $(x_0)_2$.

It is therefore useful to draw a rough graph of the function before applying Newton's method.

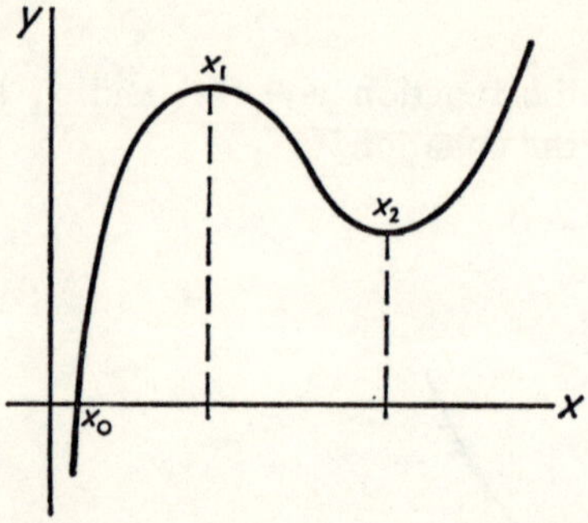

Fig. E. 2

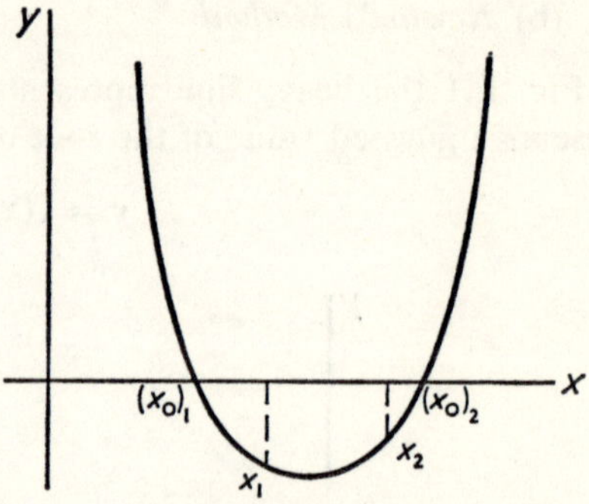

Fig. E. 3

Example. Find a root of the equation

$$2x^2+x-2 = 0$$

using Newton's method.

$$f(x) = 2x^2+x-2; \qquad f'(x) = 4x+1$$

Let $\qquad x_1 = 0{\cdot}600$

then $\qquad x_2 = 0{\cdot}600-\dfrac{2(0{\cdot}600)^2+0{\cdot}600-2}{4(0{\cdot}600)+1} = 0{\cdot}800$

$$x_3 = 0{\cdot}800-\frac{2(0{\cdot}800)^2+0{\cdot}800-2}{4(0{\cdot}800)+1} = 0{\cdot}781$$

$$x_4 = 0{\cdot}781-\frac{2(0{\cdot}781)^2+0{\cdot}781-2}{4(0{\cdot}781)+1} = 0{\cdot}781$$

The rapid convergence of x compared with the earlier method should be noted.

Logarithms

	0	1	2	3	4	5	6	7	8	9	1	2	3	4	5	6	7	8	9
10	0000	0043	0086	0128	0170	0212	0253	0294	0334	0374	4	8	12	17	21	25	29	33	37
11	0414	0453	0492	0531	0569	0607	0645	0682	0719	0755	4	8	11	15	19	23	26	30	34
12	0792	0828	0864	0899	0934	0969	1004	1038	1072	1106	3	7	10	14	17	21	24	28	31
13	1139	1173	1206	1239	1271	1303	1335	1367	1399	1430	3	6	10	13	16	19	23	26	29
14	1461	1492	1523	1553	1584	1614	1644	1673	1703	1732	3	6	9	12	15	18	21	24	27
15	1761	1790	1818	1847	1875	1903	1931	1959	1987	2014	3	6	8	11	14	17	20	22	25
16	2041	2068	2095	2122	2148	2175	2201	2227	2253	2279	3	5	8	11	13	16	18	21	24
17	2304	2330	2355	2380	2405	2430	2455	2480	2504	2529	2	5	7	10	12	15	17	20	22
18	2553	2577	2601	2625	2648	2672	2695	2718	2742	2765	2	5	7	9	12	14	16	19	21
19	2788	2810	2833	2856	2878	2900	2923	2945	2967	2989	2	4	7	9	11	13	16	18	20
20	3010	3032	3054	3075	3096	3118	3139	3160	3181	3201	2	4	6	8	11	13	15	17	19
21	3222	3243	3263	3284	3304	3324	3345	3365	3385	3404	2	4	6	8	10	12	14	16	18
22	3424	3444	3464	3483	3502	3522	3541	3560	3579	3598	2	4	6	8	10	12	14	15	17
23	3617	3636	3655	3674	3692	3711	3729	3747	3766	3784	2	4	6	7	9	11	13	15	17
24	3802	3820	3838	3856	3874	3892	3909	3927	3945	3962	2	4	5	7	9	11	12	14	16
25	3979	3997	4014	4031	4048	4065	4082	4099	4116	4133	2	3	5	7	9	10	12	14	15
26	4150	4166	4183	4200	4216	4232	4249	4265	4281	4298	2	3	5	7	8	10	11	13	15
27	4314	4330	4346	4362	4378	4393	4409	4425	4440	4456	2	3	5	6	8	9	11	13	14
28	4472	4487	4502	4518	4533	4548	4564	4579	4594	4609	2	3	5	6	8	9	11	12	14
29	4624	4639	4654	4669	4683	4698	4713	4728	4742	4757	1	3	4	6	7	9	10	12	13
30	4771	4786	4800	4814	4829	4843	4857	4871	4886	4900	1	3	4	6	7	9	10	11	13
31	4914	4928	4942	4955	4969	4983	4997	5011	5024	5038	1	3	4	5	7	8	10	11	12
32	5051	5065	5079	5092	5105	5119	5132	5145	5159	5172	1	3	4	5	7	8	9	11	12
33	5185	5198	5211	5224	5237	5250	5263	5276	5289	5302	1	3	4	5	6	8	9	10	12
34	5315	5328	5340	5353	5366	5378	5391	5403	5416	5428	1	3	4	5	6	8	9	10	11
35	5441	5453	5465	5478	5490	5502	5514	5527	5539	5551	1	2	4	5	6	7	9	10	11
36	5563	5575	5587	5599	5611	5623	5635	5647	5658	5670	1	2	4	5	6	7	8	10	11
37	5682	5694	5705	5717	5729	5740	5752	5763	5775	5786	1	2	3	5	6	7	8	9	10
38	5798	5809	5821	5832	5843	5855	5866	5877	5888	5899	1	2	3	5	6	7	8	9	10
39	5911	5922	5933	5944	5955	5966	5977	5988	5999	6010	1	2	3	4	5	7	8	9	10
40	6021	6031	6042	6053	6064	6075	6085	6096	6107	6117	1	2	3	4	5	6	7	9	10
41	6128	6138	6149	6160	6170	6180	6191	6201	6212	6222	1	2	3	4	5	6	7	8	9
42	6232	6243	6253	6263	6274	6284	6294	6304	6314	6325	1	2	3	4	5	6	7	8	9
43	6335	6345	6355	6365	6375	6385	6395	6405	6415	6425	1	2	3	4	5	6	7	8	9
44	6435	6444	6454	6464	6474	6484	6493	6503	6513	6522	1	2	3	4	5	6	7	8	9
45	6532	6542	6551	6561	6571	6580	6590	6599	6609	6618	1	2	3	4	5	6	7	8	9
46	6628	6637	6646	6656	6665	6675	6684	6693	6702	6712	1	2	3	4	5	6	7	7	8
47	6721	6730	6739	6749	6758	6767	6776	6785	6794	6803	1	2	3	4	5	5	6	7	8
48	6812	6821	6830	6839	6848	6857	6866	6875	6884	6893	1	2	3	4	4	5	6	7	8
49	6902	6911	6920	6928	6937	6946	6955	6964	6972	6981	1	2	3	4	4	5	6	7	8
50	6990	6998	7007	7016	7024	7033	7042	7050	7059	7067	1	2	3	3	4	5	6	7	8
51	7076	7084	7093	7101	7110	7118	7126	7135	7143	7152	1	2	3	3	4	5	6	7	8
52	7160	7168	7177	7185	7193	7202	7210	7218	7226	7235	1	2	2	3	4	5	6	7	7
53	7243	7251	7259	7267	7275	7284	7292	7300	7308	7316	1	2	2	3	4	5	6	6	7
54	7324	7332	7340	7348	7356	7364	7372	7380	7388	7396	1	2	2	3	4	5	6	6	7
	0	1	2	3	4	5	6	7	8	9	1	2	3	4	5	6	7	8	9

Logarithms

	0	1	2	3	4	5	6	7	8	9	1	2	3	4	5	6	7	8	9
55	7404	7412	7419	7427	7435	7443	7451	7459	7466	7474	1	2	2	3	4	5	5	6	7
56	7482	7490	7497	7505	7513	7520	7528	7536	7543	7551	1	2	2	3	4	5	5	6	7
57	7559	7566	7574	7582	7589	7597	7604	7612	7619	7627	1	2	2	3	4	5	5	6	7
58	7634	7642	7649	7657	7664	7672	7679	7686	7694	7701	1	1	2	3	4	4	5	6	7
59	7709	7716	7723	7731	7738	7745	7752	7760	7767	7774	1	1	2	3	4	4	5	6	7
60	7782	7789	7796	7803	7810	7818	7825	7832	7839	7846	1	1	2	3	4	4	5	6	6
61	7853	7860	7868	7875	7882	7889	7896	7903	7910	7917	1	1	2	3	4	4	5	6	6
62	7924	7931	7938	7945	7952	7959	7966	7973	7980	7987	1	1	2	3	3	4	5	6	6
63	7993	8000	8007	8014	8021	8028	8035	8041	8048	8055	1	1	2	3	3	4	5	6	6
64	8062	8069	8075	8082	8089	8096	8102	8109	8116	8122	1	1	2	3	3	4	5	5	6
65	8129	8136	8142	8149	8156	8162	8169	8176	8182	8189	1	1	2	3	3	4	5	5	6
66	8195	8202	8209	8215	8222	8228	8235	8241	8248	8254	1	1	2	3	3	4	5	5	6
67	8261	8267	8274	8280	8287	8293	8299	8306	8312	8319	1	1	2	3	3	4	4	5	6
68	8325	8331	8338	8344	8351	8357	8363	8370	8376	8382	1	1	2	3	3	4	4	5	6
69	8388	8395	8401	8407	8414	8420	8426	8432	8439	8445	1	1	2	3	3	4	4	5	6
70	8451	8457	8463	8470	8476	8482	8488	8494	8500	8506	1	1	2	2	3	4	4	5	6
71	8513	8519	8525	8531	8537	8543	8549	8555	8561	8567	1	1	2	2	3	4	4	5	5
72	8573	8579	8585	8591	8597	8603	8609	8615	8621	8627	1	1	2	2	3	4	4	5	5
73	8633	8639	8645	8651	8657	8663	8669	8675	8681	8686	1	1	2	2	3	4	4	5	5
74	8692	8698	8704	8710	8716	8722	8727	8733	8739	8745	1	1	2	2	3	4	4	5	5
75	8751	8756	8762	8768	8774	8779	8785	8791	8797	8802	1	1	2	2	3	3	4	5	5
76	8808	8814	8820	8825	8831	8837	8842	8848	8854	8859	1	1	2	2	3	3	4	5	5
77	8865	8871	8876	8882	8887	8893	8899	8904	8910	8915	1	1	2	2	3	3	4	4	5
78	8921	8927	8932	8938	8943	8949	8954	8960	8965	8971	1	1	2	2	3	3	4	4	5
79	8976	8982	8987	8993	8998	9004	9009	9015	9020	9025	1	1	2	2	3	3	4	4	5
80	9031	9036	9042	9047	9053	9058	9063	9069	9074	9079	1	1	2	2	3	3	4	4	5
81	9085	9090	9096	9101	9106	9112	9117	9122	9128	9133	1	1	2	2	3	3	4	4	5
82	9138	9143	9149	9154	9159	9165	9170	9175	9180	9186	1	1	2	2	3	3	4	4	5
83	9191	9196	9201	9206	9212	9217	9222	9227	9232	9238	1	1	2	2	3	3	4	4	5
84	9243	9248	9253	9258	9263	9269	9274	9279	9284	9289	1	1	2	2	3	3	4	4	5
85	9294	9299	9304	9309	9315	9320	9325	9330	9335	9340	1	1	2	2	3	3	4	4	5
86	9345	9350	9355	9360	9365	9370	9375	9380	9385	9390	1	1	2	2	3	3	4	4	5
87	9395	9400	9405	9410	9415	9420	9425	9430	9435	9440	0	1	1	2	2	3	3	4	4
88	9445	9450	9455	9460	9465	9469	9474	9479	9484	9489	0	1	1	2	2	3	3	4	4
89	9494	9499	9504	9509	9513	9518	9523	9528	9533	9538	0	1	1	2	2	3	3	4	4
90	9542	9547	9552	9557	9562	9566	9571	9576	9581	9586	0	1	1	2	2	3	3	4	4
91	9590	9595	9600	9605	9609	9614	9619	9624	9628	9633	0	1	1	2	2	3	3	4	4
92	9638	9643	9647	9652	9657	9661	9666	9671	9675	9680	0	1	1	2	2	3	3	4	4
93	9685	9689	9694	9699	9703	9708	9713	9717	9722	9727	0	1	1	2	2	3	3	4	4
94	9731	9736	9741	9745	9750	9754	9759	9764	9768	9773	0	1	1	2	2	3	3	4	4
95	9777	9782	9786	9791	9795	9800	9805	9809	9814	9818	0	1	1	2	2	3	3	4	4
96	9823	9827	9832	9836	9841	9845	9850	9854	9859	9863	0	1	1	2	2	3	3	4	4
97	9868	9872	9877	9881	9886	9890	9894	9899	9903	9908	0	1	1	2	2	3	3	4	4
98	9912	9917	9921	9926	9930	9934	9939	9943	9948	9952	0	1	1	2	2	3	3	4	4
99	9956	9961	9965	9969	9974	9978	9983	9987	9991	9996	0	1	1	2	2	3	3	4	4
	0	1	2	3	4	5	6	7	8	9	1	2	3	4	5	6	7	8	9

Anti-Logarithms

	0	1	2	3	4	5	6	7	8	9	1	2	3	4	5	6	7	8	9
·00	1000	1002	1005	1007	1009	1012	1014	1016	1019	1021	0	0	1	1	1	1	2	2	2
·01	1023	1026	1028	1030	1033	1035	1038	1040	1042	1045	0	0	1	1	1	1	2	2	2
·02	1047	1050	1052	1054	1057	1059	1062	1064	1067	1069	0	0	1	1	1	1	2	2	2
.03	1072	1074	1076	1079	1081	1084	1086	1089	1091	1094	0	0	1	1	1	1	2	2	2
·04	1096	1099	1102	1104	1107	1109	1112	1114	1117	1119	0	1	1	1	1	2	2	2	2
·05	1122	1125	1127	1130	1132	1135	1138	1140	1143	1146	0	1	1	1	1	2	2	2	2
·06	1148	1151	1153	1156	1159	1161	1164	1167	1169	1172	0	1	1	1	1	2	2	2	2
·07	1175	1178	1180	1183	1186	1189	1191	1194	1197	1199	0	1	1	1	1	2	2	2	2
·08	1202	1205	1208	1211	1213	1216	1219	1222	1225	1227	0	1	1	1	1	2	2	2	3
·09	1230	1233	1236	1239	1242	1245	1247	1250	1253	1256	0	1	1	1	1	2	2	2	3
·10	1259	1262	1265	1268	1271	1274	1276	1279	1282	1285	0	1	1	1	1	2	2	2	3
·11	1288	1291	1294	1297	1300	1303	1306	1309	1312	1315	0	1	1	1	2	2	2	2	3
·12	1318	1321	1324	1327	1330	1334	1337	1340	1343	1346	0	1	1	1	2	2	2	2	3
·13	1349	1352	1355	1358	1361	1365	1368	1371	1374	1377	0	1	1	1	2	2	2	2	3
·14	1380	1384	1387	1390	1393	1396	1400	1403	1406	1409	0	1	1	1	2	2	2	3	3
·15	1413	1416	1419	1422	1426	1429	1432	1435	1439	1442	0	1	1	1	2	2	2	3	3
·16	1445	1449	1452	1455	1459	1462	1466	1469	1472	1476	0	1	1	1	2	2	2	3	3
·17	1479	1483	1486	1489	1493	1496	1500	1503	1507	1510	0	1	1	1	2	2	2	3	3
·18	1514	1517	1521	1524	1528	1531	1535	1538	1542	1545	0	1	1	1	2	2	2	3	3
·19	1549	1552	1556	1560	1563	1567	1570	1574	1578	1581	0	1	1	1	2	2	3	3	3
·20	1585	1589	1592	1596	1600	1603	1607	1611	1614	1618	0	1	1	1	2	2	3	3	3
·21	1622	1626	1629	1633	1637	1641	1644	1648	1652	1656	0	1	1	2	2	2	3	3	3
·22	1660	1663	1667	1671	1675	1679	1683	1687	1690	1694	0	1	1	2	2	2	3	3	3
·23	1698	1702	1706	1710	1714	1718	1722	1726	1730	1734	0	1	1	2	2	2	3	3	4
·24	1738	1742	1746	1750	1754	1758	1762	1766	1770	1774	0	1	1	2	2	2	3	3	4
·25	1778	1782	1786	1791	1795	1799	1803	1807	1811	1816	0	1	1	2	2	3	3	3	4
·26	1820	1824	1828	1832	1837	1841	1845	1849	1854	1858	0	1	1	2	2	3	3	3	4
·27	1862	1866	1871	1875	1879	1884	1888	1892	1897	1901	0	1	1	2	2	3	3	3	4
·28	1905	1910	1914	1919	1923	1928	1932	1936	1941	1945	0	1	1	2	2	3	3	4	4
·29	1950	1954	1959	1963	1968	1972	1977	1982	1986	1991	0	1	1	2	2	3	3	4	4
·30	1995	2000	2004	2009	2014	2018	2023	2028	2032	2037	0	1	1	2	2	3	3	4	4
·31	2042	2046	2051	2056	2061	2065	2070	2075	2080	2084	0	1	1	2	2	3	3	4	4
·32	2089	2094	2099	2104	2109	2113	2118	2123	2128	2133	0	1	1	2	2	3	3	4	4
·33	2138	2143	2148	2153	2158	2163	2168	2173	2178	2183	0	1	1	2	2	3	3	4	4
·34	2188	2193	2198	2203	2208	2213	2218	2223	2228	2234	1	1	2	2	3	3	4	4	5
·35	2239	2244	2249	2254	2259	2265	2270	2275	2280	2286	1	1	2	2	3	3	4	4	5
·36	2291	2296	2301	2307	2312	2317	2323	2328	2333	2339	1	1	2	2	3	3	4	4	5
·37	2344	2350	2355	2360	2366	2371	2377	2382	2388	2393	1	1	2	2	3	3	4	4	5
·38	2399	2404	2410	2415	2421	2427	2432	2438	2443	2449	1	1	2	2	3	3	4	4	5
·39	2455	2460	2466	2472	2477	2483	2489	2495	2500	2506	1	1	2	2	3	3	4	5	5
·40	2512	2518	2523	2529	2535	2541	2547	2553	2559	2564	1	1	2	2	3	3	4	5	5
·41	2570	2576	2582	2588	2594	2600	2606	2612	2618	2624	1	1	2	2	3	4	4	5	5
·42	2630	2636	2642	2648	2655	2661	2667	2673	2679	2685	1	1	2	2	3	4	4	5	6
·43	2692	2698	2704	2710	2716	2723	2729	2735	2742	2748	1	1	2	2	3	4	4	5	6
·44	2754	2761	2767	2773	2780	2786	2793	2799	2805	2812	1	1	2	3	3	4	4	5	6
·45	2818	2825	2831	2838	2844	2851	2858	2864	2871	2877	1	1	2	3	3	4	5	5	6
·46	2884	2891	2897	2904	2911	2917	2924	2931	2938	2944	1	1	2	3	3	4	5	5	6
·47	2951	2958	2965	2972	2979	2985	2992	2999	3006	3013	1	1	2	3	3	4	5	6	6
·48	3020	3027	3034	3041	3048	3055	3062	3069	3076	3083	1	1	2	3	4	4	5	6	6
·49	3090	3097	3105	3112	3119	3126	3133	3141	3148	3155	1	1	2	3	4	4	5	6	7
	0	1	2	3	4	5	6	7	8	9	1	2	3	4	5	6	7	8	9

Anti-Logarithms

	0	1	2	3	4	5	6	7	8	9	1	2	3	4	5	6	7	8	9
·50	3162	3170	3177	3184	3192	3199	3206	3214	3221	3228	1	1	2	3	4	4	5	6	7
·51	3236	3243	3251	3258	3266	3273	3281	3289	3296	3304	1	2	2	3	4	5	5	6	7
·52	3311	3319	3327	3334	3342	3350	3357	3365	3373	3381	1	2	2	3	4	5	5	6	7
·53	3388	3396	3404	3412	3420	3428	3436	3443	3451	3459	1	2	2	3	4	5	6	6	7
·54	3467	3475	3483	3491	3499	3508	3516	3524	3532	3540	1	2	2	3	4	5	6	6	7
·55	3548	3556	3565	3573	3581	3589	3597	3606	3614	3622	1	2	2	3	4	5	6	7	7
·56	3631	3639	3648	3656	3664	3673	3681	3690	3698	3707	1	2	3	3	4	5	6	7	8
·57	3715	3724	3733	3741	3750	3758	3767	3776	3784	3793	1	2	3	3	4	5	6	7	8
·58	3802	3811	3819	3828	3837	3846	3855	3864	3873	3882	1	2	3	4	4	5	6	7	8
·59	3890	3899	3908	3917	3926	3936	3945	3954	3963	3972	1	2	3	4	5	5	6	7	8
·60	3981	3990	3999	4009	4018	4027	4036	4046	4055	4064	1	2	3	4	5	6	7	7	8
·61	4074	4083	4093	4102	4111	4121	4130	4140	4150	4159	1	2	3	4	5	6	7	8	9
·62	4169	4178	4188	4198	4207	4217	4227	4236	4246	4256	1	2	3	4	5	6	7	8	9
·63	4266	4276	4285	4295	4305	4315	4325	4335	4345	4355	1	2	3	4	5	6	7	8	9
·64	4365	4375	4385	4395	4406	4416	4426	4436	4446	4457	1	2	3	4	5	6	7	8	9
·65	4467	4477	4487	4498	4508	4519	4529	4539	4550	4560	1	2	3	4	5	6	7	8	9
·66	4571	4581	4592	4603	4613	4624	4634	4645	4656	4667	1	2	3	4	5	6	7	8	10
·67	4677	4688	4699	4710	4721	4732	4742	4753	4764	4775	1	2	3	4	5	7	8	9	10
·68	4786	4797	4808	4819	4831	4842	4853	4864	4875	4887	1	2	3	4	6	7	8	9	10
·69	4898	4909	4920	4932	4943	4955	4966	4977	4989	5000	1	2	3	5	6	7	8	9	10
·70	5012	5023	5035	5047	5058	5070	5082	5093	5105	5117	1	2	4	5	6	7	8	9	11
·71	5129	5140	5152	5164	5176	5188	5200	5212	5224	5236	1	2	4	5	6	7	8	10	11
·72	5248	5260	5272	5284	5297	5309	5321	5333	5346	5358	1	2	4	5	6	7	9	10	11
·73	5370	5383	5395	5408	5420	5433	5445	5458	5470	5483	1	3	4	5	6	8	9	10	11
·74	5495	5508	5521	5534	5546	5559	5572	5585	5598	5610	1	3	4	5	6	8	9	10	12
·75	5623	5636	5649	5662	5675	5689	5702	5715	5728	5741	1	3	4	5	7	8	9	10	12
·76	5754	5768	5781	5794	5808	5821	5834	5848	5861	5875	1	3	4	5	7	8	9	11	12
·77	5888	5902	5916	5929	5943	5957	5970	5984	5998	6012	1	3	4	6	7	8	10	11	12
·78	6026	6039	6053	6067	6081	6095	6109	6124	6138	6152	1	3	4	6	7	8	10	11	13
·79	6166	6180	6194	6209	6223	6237	6252	6266	6281	6295	1	3	4	6	7	9	10	12	13
·80	6310	6324	6339	6353	6368	6383	6397	6412	6427	6442	1	3	4	6	7	9	10	12	13
·81	6457	6471	6486	6501	6516	6531	6546	6561	6577	6592	2	3	5	6	8	9	11	12	14
·82	6607	6622	6637	6653	6668	6683	6699	6714	6730	6745	2	3	5	6	8	9	11	12	14
·83	6761	6776	6792	6808	6823	6839	6855	6871	6887	6902	2	3	5	6	8	9	11	13	14
·84	6918	6934	6950	6966	6982	6998	7015	7031	7047	7063	2	3	5	6	8	10	11	13	14
·85	7079	7096	7112	7129	7145	7161	7178	7194	7211	7228	2	3	5	7	8	10	12	13	15
·86	7244	7261	7278	7295	7311	7328	7345	7362	7379	7396	2	3	5	7	8	10	12	14	15
·87	7413	7430	7447	7464	7482	7499	7516	7534	7551	7568	2	3	5	7	9	10	12	14	16
·88	7586	7603	7621	7638	7656	7674	7691	7709	7727	7745	2	4	5	7	9	11	12	14	16
·89	7762	7780	7798	7816	7834	7852	7870	7889	7907	7925	2	4	5	7	9	11	13	14	16
·90	7943	7962	7980	7998	8017	8035	8054	8072	8091	8110	2	4	6	7	9	11	13	15	17
·91	8128	8147	8166	8185	8204	8222	8241	8260	8279	8299	2	4	6	8	10	11	13	15	17
·92	8318	8337	8356	8375	8395	8414	8433	8453	8472	8492	2	4	6	8	10	12	14	15	17
·93	8511	8531	8551	8570	8590	8610	8630	8650	8670	8690	2	4	6	8	10	12	14	16	18
·94	8710	8730	8750	8770	8790	8810	8831	8851	8872	8892	2	4	6	8	10	12	14	16	18
·95	8913	8933	8954	8974	8995	9016	9036	9057	9078	9099	2	4	6	8	10	12	14	17	19
·96	9120	9141	9162	9183	9204	9226	9247	9268	9290	9311	2	4	6	9	11	13	15	17	19
·97	9333	9354	9376	9397	9419	9441	9462	9484	9506	9528	2	4	7	9	11	13	15	17	20
·98	9550	9572	9594	9616	9638	9661	9683	9705	9727	9750	2	4	7	9	11	13	16	18	20
·99	9772	9795	9817	9840	9863	9886	9908	9931	9954	9977	2	5	7	9	11	14	16	18	21
	0	1	2	3	4	5	6	7	8	9	1	2	3	4	5	6	7	8	9

Trigonometrical Ratios and Radian Measure of Angles from 0° to 90° at intervals of 1°

Radians	Degrees.	Sines.	Tangents.	Secants.	Cosecants.	Cotangents.	Cosines.		
0	**0**	0	0	1	∞	∞	1	**90**	1·5708
·0175	**1**	·0175	·0175	1·0002	57·2987	57·2900	·9998	**89**	1·5533
·0349	**2**	·0349	·0349	1·0006	28·6537	28·6363	·9994	**88**	1·5359
·0524	**3**	·0523	·0524	1·0014	19·1073	19·0811	·9986	**87**	1·5184
·0698	**4**	·0698	·0699	1·0024	14·3356	14·3007	·9976	**86**	1·5010
·0873	**5**	·0872	·0875	1·0038	11·4737	11·4301	·9962	**85**	1·4835
·1047	**6**	·1045	·1051	1·0055	9·5668	9·5144	·9945	**84**	1·4661
·1222	**7**	·1219	·1228	1·0075	8·2055	8·1443	·9925	**83**	1·4486
·1396	**8**	·1392	·1405	1·0098	7·1853	7·1154	·9903	**82**	1·4312
·1571	**9**	·1564	·1584	1·0125	6·3925	6·3138	·9877	**81**	1·4137
·1745	**10**	·1736	·1763	1·0154	5·7588	5·6713	·9848	**80**	1·3963
·1920	**11**	·1908	·1944	1·0187	5·2408	5·1446	·9816	**79**	1·3788
·2094	**12**	·2079	·2126	1·0223	4·8097	4·7046	·9781	**78**	1·3614
·2269	**13**	·2250	·2309	1·0263	4·4454	4·3315	·9744	**77**	1·3439
·2443	**14**	·2419	·2493	1·0306	4·1336	4·0108	·9703	**76**	1·3265
·2618	**15**	·2588	·2679	1·0353	3·8637	3·7321	·9659	**75**	1·3090
·2793	**16**	·2756	·2867	1·0403	3·6280	3·4874	·9613	**74**	1·2915
·2967	**17**	·2924	·3057	1·0457	3·4203	3·2709	·9563	**73**	1·2741
·3142	**18**	·3090	·3249	1·0515	3·2361	3·0777	·9511	**72**	1·2566
·3316	**19**	·3256	·3443	1·0576	3·0716	2·9042	·9455	**71**	1·2392
·3491	**20**	·3420	·3640	1·0642	2·9238	2·7475	·9397	**70**	1·2217
·3665	**21**	·3584	·3839	1·0711	2·7904	2·6051	·9336	**69**	1·2043
·3840	**22**	·3746	·4040	1·0785	2·6695	2·4751	·9272	**68**	1·1868
·4014	**23**	·3907	·4245	1·0864	2·5593	2·3559	·9205	**67**	1·1694
·4189	**24**	·4067	·4452	1·0946	2·4586	2·2460	·9135	**66**	1·1519

K

·4363	**25**	·4226	·4663	1·1034	2·3662	2·1445	·9063	**65**	1·1345
·4538	**26**	·4384	·4877	1·1126	2·2812	2·0503	·8988	**64**	1·1170
·4712	**27**	·4540	·5095	1·1223	2·2027	1·9626	·8910	**63**	1·0996
·4887	**28**	·4695	·5317	1·1326	2·1301	1·8807	·8829	**62**	1·0821
·5061	**29**	·4848	·5543	1·1434	2·0627	1·8040	·8746	**61**	1·0647
·5236	**30**	·5000	·5774	1·1547	2·0000	1·7321	·8660	**60**	1·0472
·5411	**31**	·5150	·6009	1·1666	1·9416	1·6643	·8572	**59**	1·0297
·5585	**32**	·5299	·6249	1·1792	1·8871	1·6003	·8480	**58**	1·0123
·5760	**33**	·5446	·6494	1·1924	1·8361	1·5399	·8387	**57**	·9948
·5934	**34**	·5592	·6745	1·2062	1·7883	1·4826	·8290	**56**	·9774
·6109	**35**	·5736	·7002	1·2208	1·7434	1·4281	·8192	**55**	·9599
·6283	**36**	·5878	·7265	1·2361	1·7013	1·3764	·8090	**54**	·9425
·6458	**37**	·6018	·7536	1·2521	1·6616	1·3270	·7986	**53**	·9250
·6632	**38**	·6157	·7813	1·2690	1·6243	1·2799	·7880	**52**	·9076
·6807	**39**	·6293	·8098	1·2868	1·5890	1·2349	·7771	**51**	·8901
·6981	**40**	·6428	·8391	1·3054	1·5557	1·1918	·7660	**50**	·8727
·7156	**41**	·6561	·8693	1·3250	1·5243	1·1504	·7547	**49**	·8552
·7330	**42**	·6691	·9004	1·3456	1·4945	1·1106	·7431	**48**	·8378
·7505	**43**	·6820	·9325	1·3673	1·4663	1·0724	·7314	**47**	·8203
·7679	**44**	·6947	·9657	1·3902	1·4396	1·0355	·7193	**46**	·8029
·7854	**45**	·7071	1·0000	1·4142	1·4142	1·0000	·7071	**45**	·7854
		Cosines.	Cotangents.	Cosecants.	Secants.	Tangents.	Sines.	Degrees.	Radians.

The Radian Measure of any other angle can be obtained by Proportional Parts [10′ = ·0029 radian].

Table from *An Introduction to Infinitesimal Calculus* by G. W. Caunt, 1914, p. 509, by permission of The Clarendon Press, Oxford.

The Normal Distribution

This table gives the probability, *p*, of finding a value outwith *n* standard deviations from the mean.

n	*p*	*n*	*p*	*n*	*p*	*n*	*p*
0·00	1·000	0·46	0·645	0·92	0·358	1·95	0·051
0·02	0·984	0·48	0·631	0·94	0·347	2·00	0·046
0·04	0·968	0·50	0·617	0·96	0·337	2·05	0·040
0·06	0·952	0·52	0·603	0·98	0·327	2·10	0·036
0·08	0·936	0·54	0·589	1·00	0·317	2·15	0·032
0·10	0·920	0·56	0·575	1·05	0·294	2·20	0·028
0·12	0·904	0·58	0·562	1·10	0·271	2·25	0·024
0·14	0·889	0·60	0·549	1·15	0·250	2·30	0·021
0·16	0·873	0·62	0·535	1·20	0·230	2·35	0·0188
0·18	0·857	0·64	0·522	1·25	0·211	2·40	0·0164
0·20	0·841	0·66	0·509	1·30	0·194	2·45	0·0143
0·22	0·826	0·68	0·496	1·35	0·178	2·50	0·0124
0·24	0·810	0·70	0·484	1·40	0·162	2·55	0·0108
0·26	0·795	0·72	0·472	1·45	0·147	2·60	0·0093
0·28	0·779	0·74	0·459	1·50	0·134	2·65	0·0080
0·30	0·764	0·76	0·447	1·55	0·121	2·70	0·0069
0·32	0·749	0·78	0·434	1·60	0·110	2·75	0·0060
0·34	0·734	0·80	0·424	1·65	0·101	2·80	0·0051
0·36	0·719	0·82	0·412	1·70	0·089	2·85	0·0044
0·38	0·704	0·84	0·401	1·75	0·080	2·90	0·0037
0·40	0·689	0·86	0·390	1·80	0·072	2·95	0·0032
0·42	0·674	0·88	0·379	1·85	0·064	3·00	0·0027
0·44	0·660	0·90	0·368	1·90	0·057		

Table of 't' Values

Degrees of Freedom	Probability 0·20	0·10	0·05	0·01	0·0001
1	3·08	6·31	12·71	63·66	636·6
2	1·89	2·92	4·30	9·93	31·60
3	1·64	2·35	3·18	5·84	12·94
4	1·53	2·13	2·78	4·60	8·61
5	1·48	2·02	2·57	4·03	6·86
6	1·44	1·94	2·45	3·71	5·96
7	1·42	1·90	2·37	3·50	5·41
8	1·40	1·86	2·31	3·36	5·04
9	1·38	1·83	2·26	3·25	4·78
10	1·37	1·81	2·23	3·17	4·59
11	1·36	1·80	2·20	3·11	4·44
12	1·36	1·78	2·18	3·06	4·32
13	1·35	1·77	2·16	3·01	4·22
14	1·35	1·76	2·15	2·98	4·14
15	1·34	1·75	2·13	2·95	4·07
16	1·34	1·75	2·12	2·92	4·02
17	1·33	1·74	2·11	2·90	3·97
18	1·33	1·73	2·10	2·88	3·92
19	1·33	1·73	2·09	2·86	3·88
20	1·33	1·73	2·09	2·85	3·85
∞	1·28	1·64	1·96	2·58	3·29

Table of χ^2 Values

Degrees of Freedom	Probability					
	0·99	0·95	0·90	0·10	0·05	0·01
1	0·000	0·004	0·016	2·71	3·84	6·64
2	0·020	0·103	0·211	4·61	5·99	9·21
3	0·115	0·352	0·584	6·25	7·82	11·34
4	0·297	0·711	1·06	7·78	9·49	13·28
5	0·554	1·15	1·61	9·24	11·07	15·09
6	0·872	1·64	2·20	10·65	12·59	16·81
7	1·24	2·17	2·83	12·02	14·07	18·48
8	1·65	2·73	3·49	13·36	15·51	20·09
9	2·09	3·33	4·17	14·68	16·92	21·67
10	2·56	3·94	4·87	15·99	18·31	23·21
11	3·05	4·58	5·58	17·28	19·68	24·73
12	3·57	5·23	6·30	18·55	21·03	26·22
13	4·11	5·89	7·04	19·81	22·36	27·69
14	4·66	6·57	7·79	21·06	23·69	29·14
15	5·23	7·26	8·55	22·31	25·00	30·58
16	5·81	7·96	9·31	23·54	26·30	32·00
17	6·41	8·67	10·09	24·77	27·59	33·41
18	7·02	9·39	10·87	25·99	28·87	34·81
19	7·63	10·12	11·65	27·20	30·14	36·19
20	8·26	10·85	12·44	28·41	31·41	37·57

Greek alphabet

Letter		Name
Α	α	alpha
Β	β	beta
Γ	γ	gamma
Δ	δ	delta
Ε	ε	epsilon
Ζ	ζ	zeta
Η	η	eta
Θ	θ	theta
Ι	ι	iota
Κ	κ	kappa
Λ	λ	lambda
Μ	μ	mu
Ν	ν	nu
Ξ	ξ	xi
Ο	ο	omicron
Π	π	pi
Ρ	ρ	rho
Σ	σ	sigma
Τ	τ	tau
Υ	υ	upsilon
Φ	φ	phi
Χ	χ	chi
Ψ	ψ	psi
Ω	ω	omega

n	n^{-1}	$\log n!$	$n^{\frac{1}{2}}$	$(10n)^{\frac{1}{2}}$	$n^{\frac{1}{3}}$	$(10n)^{\frac{1}{3}}$	$(100n)^{\frac{1}{3}}$
1	1·00000	0·0000	1·0000	3·1623	1·0000	2·1544	4·6416
2	0·50000	0·3010	1·4142	4·4721	1·2599	2·7144	5·8480
3	0·33333	0·7782	1·7321	5·4772	1·4422	3·1072	6·6943
4	0·25000	1·3802	2·0000	6·3246	1·5874	3·4200	7·3681
5	0·20000	2·0792	2·2361	7·0711	1·7100	3·6840	7·9370
6	0·16667	2·8573	2·4495	7·7460	1·8171	3·9149	8·4343
7	0·14286	3·7024	2·6458	8·3666	1·9129	4·1213	8·8790
8	0·12500	4·6055	2·8284	8·9443	2·0000	4·3089	9·2832
9	0·11111	5·5598	3·0000	9·4868	2·0801	4·4814	9·6549
10	0·10000	6·5598	3·1623	10·0000	2·1544	4·6416	10·0000
11	0·09091	7·6012	3·3166	10·488	2·2240	4·7914	10·323
12	0·08333	8·6803	3·4641	10·955	2·2894	4·9324	10·627
13	0·07692	9·7943	3·6056	11·402	2·3513	5·0658	10·914
14	0·07143	10·940	3·7417	11·832	2·4101	5·1925	11·187
15	0·06667	12·117	3·8730	12·247	2·4662	5·3133	11·447
16	0·06250	13·321	4·0000	12·649	2·5198	5·4288	11·696
17	0·05882	14·551	4·1231	13·038	2·5713	5·5397	11·935
18	0·05556	15·806	4·2426	13·416	2·6207	5·6462	12·164
19	0·05263	17·085	4·3589	13·784	2·6684	5·7489	12·386
20	0·05000	18·386	4·4721	14·142	2·7144	5·8480	12·599
21	0·04762	19·708	4·5826	14·491	2·7589	5·9439	12·806
22	0·04546	21·051	4·6904	14·832	2·8020	6·0368	13·006
23	0·04348	22·412	4·7958	15·166	2·8439	6·1269	13·200
24	0·04167	23·793	4·8990	15·492	2·8845	6·2145	13·389
25	0·04000	25·191	5·0000	15·811	2·9240	6·2996	13·572
26	0·03846	26·606	5·0990	16·125	2·9625	6·3825	13·751
27	0·03704	28·037	5·1962	16·432	3·0000	6·4633	13·925
28	0·03571	29·484	5·2915	16·733	3·0366	6·5421	14·095
29	0·03448	30·947	5·3852	17·029	3·0723	6·6191	14·260
30	0·03333	32·424	5·4772	17·321	3·1072	6·6943	14·423

n	n^{-1}	$\log n!$	$n^{\frac{1}{2}}$	$(10n)^{\frac{1}{2}}$	$n^{\frac{1}{3}}$	$(10n)^{\frac{1}{3}}$	$(100n)^{\frac{1}{3}}$
31	0·03226	33·915	5·5678	17·607	3·1414	6·7679	14·581
32	0·03125	35·420	5·6569	17·889	3·1748	6·8399	14·736
33	0·03030	36·939	5·7446	18·166	3·2075	6·9104	14·888
34	0·02941	38·470	5·8310	18·439	3·2396	6·9795	15·037
35	0·02857	40·014	5·9161	18·708	3·2711	7·0473	15·183
36	0·02778	41·571	6·0000	18·974	3·3019	7·1138	15·326
37	0·02703	43·139	6·0828	19·235	3·3322	7·1791	15·467
38	0·02632	44·719	6·1644	19·494	3·3620	7·2432	15·605
39	0·02564	46·310	6·2450	19·748	3·3912	7·3061	15·741
40	0·02500	47·912	6·3246	20·000	3·4200	7·3681	15·874
41	0·02439	49·524	6·4031	20·248	3·4482	7·4290	16·005
42	0·02381	51·148	6·4807	20·494	3·4760	7·4889	16·134
43	0·02326	52·781	6·5574	20·736	3·5034	7·5478	16·261
44	0·02273	54·425	6·6332	20·976	3·5303	7·6059	16·386
45	0·02222	56·078	6·7082	21·213	3·5569	7·6631	16·510
46	0·02174	57·741	6·7823	21·448	3·5830	7·7194	16·631
47	0·02128	59·413	6·8557	21·680	3·6088	7·7750	16·751
48	0·02083	61·094	6·9282	21·909	3·6342	7·8297	16·869
49	0·02041	62·784	7·0000	22·136	3·6593	7·8837	16·985
50	0·02000	64·483	7·0711	22·361	3·6840	7·9370	17·100
51	0·01961	66·191	7·1414	22·583	3·7084	7·9896	17·213
52	0·01923	67·907	7·2111	22·804	3·7325	8·0415	17·325
53	0·01887	69·631	7·2801	23·022	3·7563	8·0927	17·435
54	0·01852	71·363	7·3485	23·238	3·7798	8·1433	17·544
55	0·01818	73·104	7·4162	23·452	3·8030	8·1932	17·652
56	0·01786	74·852	7·4833	23·664	3·8259	8·2426	13·758
57	0·01754	76·608	7·5498	23·875	3·8485	8·2913	17·863
58	0·01724	78·371	7·6158	24·083	3·8709	8·3396	17·967
59	0·01695	80·142	7·6811	24·290	3·8930	8·3872	18·070
60	0·01667	81·920	7·7460	24·495	3·9149	8·4343	18·171
61	0·01639	83·706	7·8102	24·698	3·9365	8·4809	18·272
62	0·01613	85·498	7·8740	24·900	3·9579	8·5270	18·371
63	0·01587	87·297	7·9373	25·100	3·9791	8·5726	18·469
64	0·01563	89·103	8·0000	25·298	4·0000	8·6177	18·566
65	0·01539	90·916	8·0623	25·495	4·0207	8·6624	18·663

n	n^{-1}	$\log n!$	$n^{\frac{1}{2}}$	$(10n)^{\frac{1}{2}}$	$n^{\frac{1}{3}}$	$(10n)^{\frac{1}{3}}$	$(100n)^{\frac{1}{3}}$
66	0·01515	92·736	8·1240	25·691	4·0412	8·7066	18·758
67	0·01493	94·562	8·1854	25·884	4·0615	8·7503	18·852
68	0·01471	96·394	8·2462	26·077	4·0817	8·7937	18·945
69	0·01449	98·233	8·3066	26·268	4·1016	8·8366	19·038
70	0·01429	100·08	8·3666	26·458	4·1213	8·8790	19·129
71	0·01409	101·93	8·4261	26·646	4·1408	8·9211	19·220
72	0·01389	103·79	8·4853	26·833	4·1602	8·9628	19·310
73	0·01370	105·65	8·5440	27·019	4·1793	9·0041	19·399
74	0·01351	107·52	8·6023	27·203	4·1983	9·0450	19·487
75	0·01333	109·39	8·6603	27·386	4·2172	9·0856	19·574
76	0·01316	111·28	8·7178	27·568	4·2358	9·1258	19·661
77	0·01299	113·16	8·7750	27·749	4·2543	9·1657	19·747
78	0·01282	115·05	8·8318	27·929	4·2727	9·2052	19·832
79	0·01266	116·95	8·8882	28·107	4·2908	9·2443	19·916
80	0·01250	118·85	8·9443	28·284	4·3089	9·2832	20·000
81	0·01235	120·76	9·0000	28·461	4·3267	9·3217	20·083
82	0·01220	122·68	9·0554	28·636	4·3445	9·3599	20·165
83	0·01205	124·60	9·1104	28·810	4·3621	9·3978	20·247
84	0·01191	126·52	9·1652	28·983	4·3795	9·4354	20·328
85	0·01177	128·45	9·2195	29·155	4·3968	9·4727	20·408
86	0·01163	130·38	9·2736	29·326	4·4140	9·5097	20·488
87	0·01149	132·32	9·3274	29·496	4·4310	9·5464	20·567
88	0·01136	134·27	9·3808	29·665	4·4480	9·5828	20·646
89	0·01124	136·22	9·4340	29·833	4·4647	9·6190	20·724
90	0·01111	138·17	9·4868	30·000	4·4814	9·6549	20·801
91	0·01099	140·13	9·5394	30·166	4·4979	9·6905	20·878
92	0·01087	142·09	9·5917	30·332	4·5144	9·7259	20·954
93	0·01075	144·06	9·6437	30·496	4·5307	9·7610	21·029
94	0·01064	146·04	9·6954	30·659	4·5468	9·7959	21·105
95	0·01053	148·01	9·7468	30·822	4·5629	9·8305	21·179
96	0·01042	150·00	9·7980	30·984	4·5789	9·8648	29·253
97	0·01031	151·98	9·8489	31·145	4·5947	9·8990	21·327
98	0·01020	153·97	9·8995	31·305	4·6104	9·9329	21·400
99	0·01010	155·97	9·9499	31·464	4·6261	9·9666	21·472
100	0·01000	157·97	10·0000	31·623	4·6416	10·0000	21·544

n	e^n	e^{-n}	sinh n	cosh n	tanh n
0·1	1·1052	0·9048	0·1002	1·0050	0·0997
0·2	1·2214	0·8187	0·2013	1·0201	0·1974
0·3	1·3499	0·7408	0·3045	1·0453	0·2913
0·4	1·4918	0·6703	0·4108	1·0811	0·3799
0·5	1·6487	0·6065	0·5211	1·1276	0·4621
0·6	1·8221	0·5488	0·6367	1·1855	0·5370
0·7	2·0138	0·4966	0·7586	1·2552	0·6044
0·8	2·2255	0·4493	0·8881	1·3374	0·6640
0·9	2·4596	0·4066	1·0265	1·4331	0·7163
1·0	2·7183	0·3679	1·1752	1·5431	0·7616
1·1	3·0042	0·3329	1·3357	1·6685	0·8005
1·2	3·3201	0·3012	1·5095	1·8107	0·8337
1·3	3·6693	0·2725	1·6984	1·9709	0·8617
1·4	4·0552	0·2466	1·9043	2·1509	0·8854
1·5	4·4817	0·2231	2·1293	2·3524	0·9051
1·6	4·9530	0·2019	2·3756	2·5775	0·9217
1·7	5·4740	0·1827	2·6456	2·8283	0·9354
1·8	6·0497	0·1653	2·9422	3·1075	0·9468
1·9	6·6859	0·1496	3·2682	3·4177	0·9562
2·0	7·3891	0·1353	3·6269	3·7622	0·9640
2·1	8·1662	0·1225	4·0219	4·1443	0·9705
2·2	9·0250	0·1108	4·4571	4·5679	0·9757
2·3	9·9742	0·1003	4·9370	5·0372	0·9801
2·4	11·023	0·0907	5·4662	5·5570	0·9837
2·5	12·183	0·0821	6·0502	6·1323	0·9866
2·6	13·464	0·0743	6·6947	6·7690	0·9890
2·7	14·880	0·0672	7·4063	7·4735	0·9910
2·8	16·445	0·0608	8·1919	8·2527	0·9926
2·9	18·174	0·0550	9·0596	9·1146	0·9940
3·0	20·086	0·0498	10·018	10·068	0·9951

Element	Symbol	Atomic Number	Atomic Weight	Element	Symbol	Atomic Number	Atomic Weight
Actinium	Ac	89	(227)	Erbium	Er	68	167·26
Aluminium	Al	13	26·98	Europium	Eu	63	151·96
Americium	Am	95	(243)	Fermium	Fm	100	(257)
Antimony	Sb	51	121·75	Fluorine	F	9	19·00
Argon	Ar	18	39·95	Francium	Fr	87	(223)
Arsenic	As	33	74·92	Gadolinium	Gd	64	157·25
Astatine	At	85	(210)	Gallium	Ga	31	69·72
Barium	Ba	56	137·34	Germanium	Ge	32	72·59
Berkelium	Bk	97	(247)	Gold	Au	79	196·97
Beryllium	Be	4	9·01	Hafnium	Hf	72	178·49
Bismuth	Bi	83	208·98	Helium	He	2	4·00
Boron	B	5	10·81	Holmium	Ho	67	164·93
Bromine	Br	35	79·90	Hydrogen	H	1	1·01
Cadmium	Cd	48	112·40	Indium	In	49	114·82
Calcium	Ca	20	40·08	Iodine	I	53	126·90
Californium	Cf	98	(251)	Iridium	Ir	77	192·22
Carbon	C	6	12·01	Iron	Fe	26	55·85
Cerium	Ce	58	140·12	Krypton	Kr	36	83·80
Caesium	Cs	55	132·91	Lanthanum	La	57	138·91
Chlorine	Cl	17	35·45	Lawrencium	Lw	103	(257)
Chromium	Cr	24	52·00	Lead	Pb	82	207·19
Cobalt	Co	27	58·93	Lithium	Li	3	6·94
Copper	Cu	29	63·55	Lutetium	Lu	71	174·97
Curium	Cm	96	(247)	Magnesium	Mg	12	24·31
Dysprosium	Dy	66	162·50	Manganese	Mn	25	54·94
Einsteinium	Es	99	(254)	Mendelevium	Md	101	(256)

Element	Symbol	Atomic Number	Atomic Weight	Element	Symbol	Atomic Number	Atomic Weight
Mercury	Hg	80	200·59	Samarium	Sm	62	150·35
Molybdenum	Mo	42	95·94	Scandium	Sc	21	44·96
Neodymium	Nd	60	144·24	Selenium	Se	34	78·96
Neon	Ne	10	20·18	Silicon	Si	14	28·09
Neptunium	Np	93	237·05	Silver	Ag	47	107·87
Nickel	Ni	28	58·71	Sodium	Na	11	22·99
Niobium	Nb	41	92·91	Strontium	Sr	38	87·62
Nitrogen	N	7	14·01	Sulphur	S	16	32·06
Nobelium	No	102	(254)	Tantalum	Ta	73	180·95
Osmium	Os	76	190·2	Technetium	Tc	43	(98·91)
Oxygen	O	8	16·00	Tellurium	Te	52	127·60
Palladium	Pd	46	106·4	Terbium	Tb	65	158·93
Phosphorus	P	15	30·97	Thallium	Tl	81	204·37
Platinum	Pt	78	195·09	Thorium	Th	90	232·04
Plutonium	Pu	94	(244)	Thulium	Tm	69	168·93
Polonium	Po	84	(210)	Tin	Sn	50	118·69
Potassium	K	19	39·10	Titanium	Ti	22	47·90
Praseodymium	Pr	59	140·91	Tungsten	W	74	183·85
Prometheum	Pm	61	(145)	Uranium	U	92	238·03
Protoactinium	Pa	91	231·04	Vanadium	V	23	50·94
Radium	Ra	88	226·03	Xenon	Xe	54	131·30
Radon	Rn	86	(222)	Ytterbium	Yb	70	173·04
Rhenium	Re	75	186·2	Yttrium	Y	39	88·91
Rhodium	Rh	45	102·91	Zinc	Zn	30	65·37
Rubidium	Rb	37	85·47	Zirconium	Zr	40	91·22
Ruthenium	Ru	44	101·07				

CONSTANTS

	c.g.s.	S.I.
Velocity of light in vacuum	$2{\cdot}998 \times 10^{10}$ cm s^{-1}	$2{\cdot}998 \times 10^{8}$ m s^{-1}
Mass of electron	$9{\cdot}110 \times 10^{-28}$ g	$9{\cdot}110 \times 10^{-31}$ kg
Mass of neutron	$1{\cdot}675 \times 10^{-24}$ g	$1{\cdot}675 \times 10^{-27}$ kg
Mass of proton	$1{\cdot}673 \times 10^{-24}$ g	$1{\cdot}673 \times 10^{-27}$ kg
Charge on electron	$4{\cdot}803 \times 10^{-10}$ esu (statcoulomb)	$1{\cdot}602 \times 10^{-19}$ C
Charge to mass ratio for electron	$5{\cdot}273 \times 10^{17}$ esu g^{-1}	$1{\cdot}759 \times 10^{11}$ C kg^{-1}
Permittivity of vacuum (ε_0)	$8{\cdot}847 \times 10^{-14}$ F cm^{-1}	$8{\cdot}847 \times 10^{-12}$ F m^{-1}
Permeability of vacuum (μ_0)	$1{\cdot}257 \times 10^{-8}$ H cm^{-1}	$1{\cdot}257 \times 10^{-6}$ H m^{-1}
Avogadro's number (N_0)	$6{\cdot}022 \times 10^{23}$ molecules mole^{-1}	$6{\cdot}022 \times 10^{23}$ molecules mol^{-1}
Planck's constant (h)	$6{\cdot}626 \times 10^{-27}$ erg s	$6{\cdot}626 \times 10^{-34}$ J s
Boltzmann's constant (k)	$1{\cdot}381 \times 10^{-16}$ erg deg^{-1} molecule^{-1}	$1{\cdot}381 \times 10^{-23}$ J K^{-1} molecule^{-1}
Faraday's constant (F)	$9{\cdot}649 \times 10^{4}$ C equiv^{-1}	$9{\cdot}649 \times 10^{4}$ C equiv^{-1}
Rydberg's constant (R_∞)	$1{\cdot}097 \times 10^{5}$ cm^{-1}	$1{\cdot}097 \times 10^{7}$ m^{-1}
Gas constant (R)	$8{\cdot}314 \times 10^{7}$ erg deg^{-1} mole^{-1}	$8{\cdot}314$ J K^{-1} mol^{-1}
	$1{\cdot}987$ cal deg^{-1} mole^{-1}	
	$8{\cdot}206 \times 10^{-2}$ l atm deg^{-1} mole^{-1}	
2·303 RT/F	$5{\cdot}916 \times 10^{-2}$ V (at 25°C)	$5{\cdot}916 \times 10^{-2}$ V (at 25°C)
	$5{\cdot}817 \times 10^{-2}$ V (at 20°C)	$5{\cdot}817 \times 10^{-2}$ V (at 20°C)
Molar volume for ideal gas at STP	$2{\cdot}241 \times 10^{4}$ cm^{3} mole^{-1}	$2{\cdot}241 \times 10^{-2}$ m^{3} mol^{-1}

π	3·141 59
e	2·718 28
$\log_e 10$	2·302 58

CONVERSION FACTORS

Angle	1 radian	57·296°
	1°	0·017 45 radians
Frequency	1 cycle s^{-1}	1·000 Hz
	1 fresnel	$1{\cdot}000 \times 10^{12}$ Hz
Length	1 angstrom (Å)	$1{\cdot}000 \times 10^{-10}$ m
	1 atomic unit (a.u.)	$5{\cdot}292 \times 10^{-11}$ m
	1 micron (μ)	$1{\cdot}000 \times 10^{-6}$ m
Force	1 dyne	$1{\cdot}000 \times 10^{-5}$ N
Pressure	1 atm	$1{\cdot}013 \times 10^{5}$ N m^{-2}
	1 torr (1 mm Hg)	$1{\cdot}333 \times 10^{2}$ N m^{-2}
	1 bar	$1{\cdot}000 \times 10^{5}$ N m^{-2}
Energy	1 erg	$1{\cdot}000 \times 10^{-7}$ J
	1 cal	4·184 J
	1 eV	$1{\cdot}602 \times 10^{-19}$ J
	1 eV $molecule^{-1}$	$9{\cdot}646 \times 10^{4}$ J mol^{-1}
	1 atomic unit (a.u.)	$4{\cdot}354 \times 10^{-18}$ J
Charge	1 esu (statcoulomb)	$3{\cdot}333 \times 10^{-10}$ C
Magnetic moment	1 Bohr magneton ($9{\cdot}274 \times 10^{-21}$ erg $gauss^{-1}$)	$9{\cdot}274 \times 10^{-24}$ A m^2
	1 nuclear magneton ($5{\cdot}050 \times 10^{-24}$ erg $gauss^{-1}$)	$5{\cdot}051 \times 10^{-27}$ A m^2
Mass	1 atomic unit (a.u.)	$1{\cdot}661 \times 10^{-27}$ kg

Index